FALLING BEHIND: INTERNATIONAL SCRUTINY OF THE PEACEFUL ATOM

Henry D. Sokolski
Editor

February 2008

*****

*****

Comments pertaining to this report are invited and should be forwarded to: Director, Strategic Studies Institute, U.S. Army War College, 122 Forbes Ave, Carlisle, PA 17013-5244.

*****

All Strategic Studies Institute (SSI) publications are available on the SSI homepage for electronic dissemination. Hard copies of this report also may be ordered from our homepage. SSI's homepage address is: *www.StrategicStudiesInstitute.army.mil.*

*****

The Strategic Studies Institute publishes a monthly e-mail newsletter to update the national security community on the research of our analysts, recent and forthcoming publications, and upcoming conferences sponsored by the Institute. Each newsletter also provides a strategic commentary by one of our research analysts. If you are interested in receiving this newsletter, please subscribe on our homepage at *www.StrategicStudiesInstitute.army.mil/newsletter/.*

ISBN 1-58487-339-6

CONTENTS

FOREWORD

The following volume consists of research that The Nonproliferation Policy Education Center (NPEC) commissioned in 2005 and 2006. This work was critiqued at a set of international conferences held in London, England, at King's College; Washington, DC; and in Paris, France, in cooperation with the *Fondation pour la Recherche Stratégique* (FRS) and the French Foreign Ministry. Dr. Wyn Bowen at King's College, Bruno Gruselle at FRS, and Martin Briens of the French Foreign Ministry were critical to the success of these meetings.

Funding for the project came from the Carnegie Corporation of New York. Both of the corporation's key project managers, Stephen Del Rosso and Patricia Moore Nicholas, were generous with their time and always supportive. NPEC's project coordinator, Tamara Mitchell, and the Strategic Studies Institute (SSI) staff helped prepare the book manuscript. Without Ms. Mitchell's help and that of Ms. Marianne Cowling and Ms. Rita Rummel of SSI, the book would not have been possible. Finally, to the project's authors and participants who contributed their time, ideas, and hard work, a special thanks is due.

HENRY D. SOKOLSKI
Executive Director
The Nonproliferation Policy
Education Center

PART I:
INTRODUCTION AND OVERVIEW

CHAPTER 1

ASSESSING THE IAEA'S ABILITY TO VERIFY THE NPT

A Report of the Nonproliferation Policy Education Center on the International Atomic Energy Agency's Nuclear Safeguards System

Henry D. Sokolski

OVERVIEW

Ask how effective International Atomic Energy Agency (IAEA) nuclear safeguards are in blocking proliferation, and you are sure to get a set of predictable reactions. Those skeptical of the system will complain that IAEA inspections are too sketchy to ferret out nuclear misbehavior (e.g., North Korea, Iraq, and Iran) and that in the rare cases when such violators are found out (almost always by national intelligence agencies), the IAEA's board of governors is loath to act. IAEA supporters have a rather opposite view. The IAEA, they point out, actually found Pyongyang, Baghdad, and Tehran in non-compliance with their IAEA safeguards agreements and reported this to the United Nations (UN) Security Council. International inspectors, moreover, were the only ones correctly to assess the status of Saddam's strategic weapons programs. The problem is not to be found in Vienna or in the IAEA's inspections system but in Washington's unwillingness to listen. In the future, the United States, they argue, should rely more, not less, on the IAEA to sort out Iran's nuclear activities and to disable North Korea's nuclear weapons complex.

These two views could hardly be more opposed. There is at least one point, though, upon which both sides agree: If possible, it would be useful to enhance the IAEA's ability to detect and prevent nuclear diversions. This would not only reduce the current risk of nuclear proliferation, it would make the further expansion of nuclear power much less risky.

The question is what is possible? To date, little has been attempted to answer this basic question. Periodic reports by the U.S. Government Accountability Office (GAO) and the IAEA have highlighted budgetary, personnel, and and administrative challenges that are immediately facing the agency.[1] There also has been a 2-year internal IAEA review of how existing IAEA safeguards procedures might be improved.[2] None of these assessments, however, has tackled the more fundamental question of how well the IAEA is actually doing in achieving its nuclear material accountancy mission. Precisely what nuclear activities and materials can the IAEA monitor to detect a diversion early enough to prevent it? What inherent limits does the IAEA nuclear inspections system face? In light of these limits, what new initiatives should the IAEA Department of Safeguards attempt and, even more important, stay clear of? What additional authority and technical capabilities might the IAEA secure to help achieve its nuclear material accountancy goals? In the end, what is or should be protected as being "peaceful" under the Nuclear Nonproliferation Treaty (NPT) or the IAEA charter? What is the proper balance between expanding the use of nuclear energy and making sure it is not diverted to make bombs?

None of these questions admits to quick or easy answers. All, however, are increasingly timely. Will IAEA safeguards be able keep Iran from using their nuclear programs to make bombs? What of IAEA's inspectors' abilities to ferret out all of North Korea's nuclear activities? Will the safeguards being proposed for India effectively prevent U.S. and foreign nuclear cooperation from assisting New Delhi's nuclear weapons program?

Then, there is the long-term problem of nuclear power's possible expansion. Since 2005, more than fifteen countries have announced a desire to acquire large reactors of their own by 2020 (this is on top of the 31 nations that already operate such reactors).[3] Nine of these states—Algeria, Morocco, Tunisia, Libya, Egypt, Turkey, Jordan, Saudi Arabia, and Yemen—are located in the war-torn region of the Middle East. Morocco, Tunisia, Jordan, and Yemen seem unlikely to achieve their stated goal. But the others, with U.S., Chinese, French and Russian nuclear cooperation, may well succeed. What is clear is that most are interested in developing a nuclear program capable of more than merely boiling water to run turbines that generate electricity. At least four have made it clear that they are interested in hedging their security bets with a nuclear weapons-option. For these states, developing purportedly peaceful nuclear energy is the weapon of choice. Will the IAEA, which is pledged to keep these programs peaceful, be able to do so?

In anticipation of these nuclear challenges, the Nonproliferation Policy Education Center (NPEC) began in 2005 to consult with officials from the IAEA, the United States, the United Kingdom, the United Arab Emirates, Germany, and France, as well as outside experts on the effectiveness of the IAEA's safeguards system and how best to improve it. NPEC went on to commission 13 studies on a variety of safeguards-related issues. These analyses were reviewed and discussed at a series of private conferences with senior level officials and outside experts held in Washington, Paris, and London.[4]

A key conclusion of these meetings and research was that the IAEA is already falling behind in achieving its material accountancy mission and risks slipping further unless members of the IAEA board independently and in concert take remedial actions in the next 2 to 5 years. The most important of these measures can be organized around seven basic recommendations:

1. ***Resist calls to read the NPT as recognizing the per se right to any and all nuclear technology, no matter how unsafeguardable or uneconomic such technology might be.*** The current, permissive, mistaken interpretation of the NPT is that all states have a sovereign *per se* right to any and all nuclear technology and materials, including nuclear fuel making and nuclear weapons usable materials, so long as they are declared to the IAEA, occasionally inspected, and have some conceivable civilian application. This interpretation, if not overturned, will guarantee a world full of nuclear weapons-ready states. With only a few more such states, the IAEA's ability to detect military diversions in a timely fashion will be marginal at best. For this reason, as well as a series of legal, historical, and technical reasons, it is essential that members of the IAEA Board of Governors make the IAEA's ability to detect military nuclear diversions in a timely fashion and the economic viability of any nuclear project to be two clear criteria for what is peaceful and protected under the NPT. Nuclear power also should only be considered to be peaceful and beneficial if it makes at least as much economic sense as its nonnuclear alternatives. Thirty years ago, the United States stipulated that in Title V of the Nuclear Nonproliferation Act of 1978 (see Title V, The Nuclear Nonproliferation Act of 1978, P.L. 95-242) Sections 501-503) that the U.S. executive branch should create a series of international technical cooperative programs

to promote the use of non-nuclear and non-petroleum renewable energy sources. The law also required the executive branch to conduct country-specific energy assessments and to report annually on the progress of U.S. and international efforts to employ such energy sources abroad. Unfortunately, since the law's passage, the White House and the U.S. Departments of Energy and State have yet to comply with any of the legal requirements of this title.

Specific Recommendations:

A. The United States and like-minded nations should stipulate in the run up to the 2010 NPT Review Conference that future civilian nuclear energy projects should only enjoy the protection of the NPT if they are:

(1) able to be monitored in non-nuclear-weapon states so as to afford timely warning of military diversions as stipulated by the NPT and the IAEA's own official criteria for what effective safeguards require; and,

(2) economically viable enough to be financed *without* nuclear-specific government subsidies.

B. The U.S. Government should begin full implementation of Section V of the Nuclear Nonproliferation Act of 1978 and urge its closest allies to cooperate in achieving its stated goals.

2. ***Distinguish between what actually can be effectively safeguarded, and what can, at best, only be monitored.*** Currently, the IAEA is unable to provide timely warning of diversions from nuclear fuel-making plants (enrichment, reprocessing, and fuel processing plants utilizing nuclear materials directly useable to make bombs). For some of these plants, the agency loses track of many nuclear weapons-worth of material every year. Meanwhile, the IAEA is unable

to prevent the overnight conversion of centrifuge enrichment and plutonium reprocessing plants into nuclear bomb-material factories. As the number of these facilities increases, the ability of the agency to fulfill its material accountancy mission dangerously erodes. The IAEA has yet to concede these points by admitting that although it can monitor these dangerous nuclear activities, it cannot actually do so in a manner that can assure *timely detection* of a possible military diversion—the key to an inspection procedure being a safeguard against military diversions. In addition, the IAEA's original criteria for how much nuclear material is needed to make one bomb (a "significant quantity"), for how much time is required to convert various materials into bombs ("conversion time"), and what the IAEA's own inspection goals should consequently be ("timeliness detection goals") were set over 30 years ago and need updating.

Specific Recommendations:

A. Require the IAEA Department of Safeguards to distinguish between those nuclear activities and materials for which timely detection of military diversions is actually possible and those for which it is not possible. This could be encouraged by having the nuclear weapons state members of the IAEA do their own individual, national analyses of these questions and make their findings public.

B. In light of the nuclear inspections experience of the last 15 years with North Korea, Iraq, Iran, Egypt, Taiwan, Libya, and South Korea, members of the IAEA Board of Governors should be encouraged to undertake their own national reassessment of what the IAEA's current significant quantities criteria, conversion times, and timely detection goals should be. These reassessments would be driven by what the

IAEA would need to assure timely detection of military diversions—i.e., time sufficient to allow states to intervene to block the possible high-jacking of civilian facilities and materials to make bombs. On the basis of these analyses, the IAEA Board of Governors should instruct the IAEA Department of Safeguards to report back to the Board regarding desirable revisions to the agency's criteria for what nuclear safeguards over different nuclear materials and activities should be.

C. Call for increased monitoring of those nuclear facilities for which such timely detection is not yet possible (e.g., nuclear bulk-handling facilities where nuclear fuel is made and processed and on-line fueled reactors, such as heavy water reactors, where keeping track of the fuel going in an out of the plant is particularly taxing). Such increased monitoring should be designed at least to increase the prospect of detecting diversions *after* they have occurred. The IAEA should make clear that timely detection of diversions (i.e., detection of diversions *before* they are completed) from such facilities is not yet possible. Finally, the IAEA should make the plant operators and owners pay for this additional monitoring. This additional cost should be considered a normal cost of conducting these activities.

D. Avoid involving the IAEA in the verification of a military fissile material cut-off treaty (FMCT). As currently proposed, a FMCT assumes that the timely detection of diversions from declared nuclear fuel-making plants is possible when, in fact, it clearly is not.

E. Call for physical security measures at those facilities where timely detection is not possible that are equivalent to the most stringent standards currently employed in nuclear-weapons facilities in the United

States, Britain, Russia, China, and France. Again, the cost of such additional security measures should be born by the owner or operator.

3. ***Reestablish material accountancy as the IAEA's top safeguards mission*** by pacing the size and growth in the agency's safeguards budget against the size and growth of the number of significant quantities of special material and bulk handling facilities that the agency must account for and inspect (see Figure 1, p. 20 below). As noted above, the amounts of special nuclear material under IAEA safeguards that go unaccounted for is large and increasing every year. These increases are most worrisome in non-weapons states that are now making nuclear fuel (e.g., Iran, Japan, the Netherlands, Germany, and Brazil). Unfortunately, the IAEA refuses to report anything but aggregate information about these materials: There are no national breakdowns that are publicly available for the different types of nuclear fuels being safeguarded in each country nor a run down of the materials that have gone unaccounted for country-by-country. As already noted, the IAEA is technically unable to meet its own timely detection goals for the safeguarding of plants producing and processing separated plutonium, highly enriched uranium, and mixed oxide fuels. Candor and encouraging restraint is all that can currently be offered to address this safeguards gap. In addition, at most of the sites that it must safeguard, the IAEA lacks the near-real time monitoring capabilities necessary to determine if the agency's own monitoring cameras and other sensors (which are left unattended for 90 or more days) are actually turned on. As such, a proliferator could divert entire fuel rods containing one or more significant quantities of lightly enriched

uranium and nuclear weapons-usable plutonium without the agency finding out either at all or in a timely fashion. Unlike the safeguards gap associated with nuclear fuel producing and processing plants, though, this gap can technically be fixed by installing near-real time surveillance systems that allow IAEA inspectors in Vienna to receive information from the remote sensors it has deployed without being on site. Certainly before the IAEA takes on additional dubious or extremely challenging missions, such as monitoring fissile production cut-offs or searching for nuclear weapons-related activities, it must arrest this growing gap between the amounts of nuclear materials it must safeguard and its technical ability to do so.

Specific Recommendations:

A. Pay greater attention to what the IAEA can clearly do better—count fresh and spent fuel rods—by quickly increasing and optimizing its remote near-real time monitoring capabilities for all of its monitoring systems, and increasing the number of full-time, qualified nuclear inspectors necessary to conduct on-site inspections.

B. Require the IAEA Department of Safeguards to report annually to the public on its safeguards budget and identify not only the number of man-hours dedicated to onsite inspections and the number of significant quantities under the IAEA's safeguards charge, but also the amount of direct-use materials (materials that can be quickly turned into bomb fuel) under its charge by type *for which the agency could not achieve its own timeliness detection goals*, the amount of direct-use materials for which the agency could achieve its own timeliness detection goals; the number and location of facilities under near-real time surveillance; the amount of money dedicated to wide-area surveillance; and the amount of money dedicated to IAEA safeguards research and development. In each case, the IAEA should present national breakdowns of each total.

C. In addition, each member state of the IAEA Board of Governors should routinely conduct its own national analysis of what it believes the proper ways to the address the problems noted above are and publicly identify and explain what it thinks the agency's top safeguards priority should be to improve these numbers.

4. ***Focus greater attention on useful safeguards activities that are necessary, but have yet to be fully developed.*** To assure that the IAEA's material accountancy assets do not risk becoming cannibalized for other urgent missions that might arise (e.g., inspections for India if the U.S.-India nuclear deal should go forward, more intrusive inspections for Iran, and North Korea, etc.), it would be useful for the agency to develop stand-by wide-area surveillance teams for the imposition of sudden inspections requirements. The agency might also usefully do more to account for source materials in processed form, as it was information regarding the shipment of such material from China that originally tipped off the IAEA to suspicious nuclear activities in Iran. The agency also needs fully to fund and properly staff its sampling analysis facilities and its efforts to secure overhead imagery of the sites that it must inspect. Finally, the agency needs to do more to establish what its own safeguards research and development requirements might be.

Specific Recommendations:

A. Members of the IAEA Board of Governors should assess on their own what might be required to conduct wide-area surveillance inspections of Iran and North Korea (i.e., what such inspections would cost to stand up and maintain in terms of dollars and staff), and ask the IAEA Board of Governors to task the IAEA Department of Safeguards to do likewise.

B. The IAEA Board of Governors should ask its members for supplemental contributions to stand up and maintain such surveillance units so that they can be tapped at any time without affecting the IAEA's routine safeguards operations. To the extent possible, the supplemental contributions should be based on a formula tied to the costs of generating nuclear electricity in each member state (as called for by recommendation 5 detailed below).

C. Similar studies should be conducted and supplemental assessments made in support of IAEA efforts to improve the agency's ability to account for nuclear source material and to fund nuclear sampling analyses and of inspections-related overhead imagery and analysis.

5. ***Complement the existing UN formula for raising IAEA funding with a user-fee for safeguards paid for by each nuclear operator.*** The IAEA's director general has repeatedly noted how small the agency's safeguards budget is, but has yet to propose how to increase it. As a stop-gap measure, the United States, European Union (EU), and Japan have been giving token amounts of voluntary, "supplemental" contributions to the agency. Currently, the UN formula used to raise IAEA funds has nations that possess no power reactors, such as Italy, paying more than nations, such

as South Korea, that possess 20 such plants. Countries including the United States, Canada, Brazil, Japan, and India, meanwhile, are taxing the IAEA safeguards system (or soon will be) with nuclear fuel-making and bulk-handling facilities and on-line fueled reactors that are much more challenging to monitor than other nuclear plants. Although the IAEA inspects the nuclear reactors and facilities of nuclear-weapon state members of the NPT far less than they inspect those of the non-nuclear-weapon states, the nuclear-weapon NPT states arguably have the most to gain from IAEA efforts to prevent the further spread of nuclear weapons. Both the insufficiency of the IAEA safeguards spending and the inequity of the way funds are currently raised for this function suggest the need to complement existing country assessments with a safeguards surcharge that is based on the costs of generating nuclear energy in each country. This surcharge is needed to assure the IAEA's budget not only grows significantly above its current level (which is too low by one or two orders of magnitude), but also to keep up with the possible expansion of nuclear power.

Specific Recommendations:

A. The United States, EU and Japan each should base all of their current supplemental contributions to the IAEA safeguards budget on a national formula based on a specific percentage of nuclear generating costs as it relates to the number of kilowatt hours that their civilian reactors generate per year.

B. The United States, EU, and Japan should, then, negotiate among themselves on what an agreed safeguards surcharge formula should be and encourage others to follow suit so that revenues from such a fee would become mandatory for each country contributing to the IAEA and would go exclusively to support the IAEA's Department of Safeguards. The UN formula, meanwhile, would be used to support the IAEA's non-safeguarding activities.

C. The IAEA Board of Governors should instruct the agency's Department of Safeguards to identify those nuclear facilities (e.g., on-line fueled reactors and nuclear fuel making plants) that require the greatest amount of resources to inspect or pose the greatest difficulty in meeting the agency's own timely detection criteria. The IAEA Board of Governors should then ask those countries possessing these identified facilities to pay an additional amount to the IAEA Department of Safeguards to cover the additional costs associated with their inspection. To the extent possible, the IAEA should encourage nations having to pay such additional fees to collect them from the customers or owners or operators of these facilities.

6. ***Establish default actions against various levels of IAEA safeguards agreement non-compliance.*** Currently, any proliferator that violates its IAEA comprehensive safeguards agreement knows that the deck is stacked against the IAEA Board of Governors reaching a consensus to (1) find them in non-compliance, and (2) take any disciplinary action. The key reason why is simple: The current burden of proof regarding any non-compliance issue is on the IAEA staff and the Board of Governors rather than on the suspect proliferator. In the absence of political consensus in the IAEA Board of Governors, the proliferator can be assured that no non-compliance finding will be made, much less any disciplinary action taken. This set of operating assumptions needs to be reversed. The best way to assure this is to establish a set of country-neutral rules regarding non-compliance that will go into effect

automatically upon the Board of Governors' inability to reach a consensus on (1) whether or not a given party is in full compliance with its comprehensive safeguards agreement, and (2) what action to take in the event that a party is found to be in non-compliance.

Specific Recommendations:

A. The United States, EU, and other like-minded nations should announce—independent of Nuclear Suppliers Group (NSG) consensus—that they will suspend transfers of controlled nuclear goods from their jurisdiction to any country that the IAEA Board of Governors has been unable to find in full compliance with its safeguards obligations and urge the IAEA Board of Governors and the NSG to agree to do the same. Under such a regime, the IAEA Board of Governors would be forced to suspend nuclear cooperation from any IAEA member to the suspected state until the Board could unanimously determine that the suspect state was in full compliance.

B. The United States, EU, and other like-minded nations should call on the UN Security Council (UNSC) to pass resolutions prohibiting states found in non-compliance by the IAEA Board of Governors from making nuclear fuel for a decade, and requiring these states to submit to intrusive wide-area surveillance to establish that they are completely out of the bomb-making business.

C. The United States, individual EU member states, and other like-minded nations should take national actions to sanction states that withdraw from the NPT while in violation of the treaty and call on the UNSC, IAEA and the NSG to pass a country-neutral sanctions resolution that tracks these sanctions measures.

D. At a minimum, the United States and like-minded states should adopt national laws and executive orders to establish what sanctions they would be willing to impose against any non-nuclear-weapon state that tests a nuclear device and call on the UNSC to pass a country-neutral sanctions resolution that tracks these national sanctions. The sanctions could be lifted after the testing state has disarmed and demonstrated to the IAEA Board of Governors that they are out of the bomb making business.

7. ***Plan on meeting future safeguards requirements on the assumption that the most popular innovations—integrated safeguards, "proliferation-resistant" fuel-cycles, and international fuel assurances—may not achieve their stated goals or, worse, may undermine them.*** Perhaps the three most popular safeguards innovations—integrated safeguards under the Additional Protocol, proliferation-resistant fuel-cycles under America's Global Nuclear Energy Partnership (GNEP), and international fuel assurances that can be afforded through fuel banks and regional fuel-making centers—are also the most unexamined. Recent analyses conducted by outside think-tanks (including the Council on Foreign Relations, Princeton University's International Panel on Fissile Materials, the Keystone Center, and the U.S. National Laboratories), in fact, conclude that each of these innovations could prove to be ineffectual or even self-defeating. GNEP's proposed proliferation-resistant fuel-cycles, for example, do not appear to be very proliferation resistant especially with respect to state-based proliferation and could easily increase the use and availability of nuclear weapons-usable fuels worldwide. Fuel banks and fuel making centers, if they make fuel available at "affordable" or "reasonable" prices, could easily end up subsidizing nuclear power development in regions where such

activity would not be economical or safe. Fuel-making centers also could end up spreading nuclear-fuel making technology. Finally, integrated safeguards, which reduce the number of inspections per safeguarded facility, could easily become a crutch for the IAEA to evade its material accountancy responsibilities.

Specific Recommendation: The U.S. Government should create a board of outside experts to serve as a quality-assurance panel to spot the potential downsides of any nonproliferation initiative. This group would be created by and report to Congress on the potential self-defeating consequences of any proposed government "nonproliferation" initiative might have prior to Congress authorizing or appropriating to support it.

Some of these recommendations are easy to act upon; others are not. IAEA member states, though, should begin to act on them now. Certainly, it would be a mistake to wait to see if civilian nuclear energy will expand (a proposition whose demonstration may require another decade or more). The reason why is simple: Even if nuclear power does not expand, the amount of nuclear weapons-usable materials that the IAEA must prevent from being diverted to make bombs is already very large and growing.

SOME NEGATIVE TRENDS

On a number of counts, the IAEA safeguards system appears to be getting better. After more than a decade of no real growth, annual funding for nuclear inspections finally was increased in real terms from $89 million in 2003 to $102 million in 2004 and to $108 million in 2007. Deployment of advanced remote monitoring equipment is on the rise and implementation of new, more intrusive inspections authority under the Additional Protocol is moving forward. In the future, nuclear power might expand, but most of this expansion will take place in nuclear weapons states or countries that are so trustworthy that it could be argued that few, if any, additional nuclear inspections may be needed. As for additional safeguards requirements—e.g., inspections in India, North Korea, or Iran—they might well be met with additional contributions when and if they arise. From this perspective, current safeguards budgeting and planning could be viewed as being adequate to the task for years to come.[5]

It could, that is, until other, less positive trends are considered. Of these, perhaps the most important concerns the number of significant quantities of nuclear material that the IAEA must safeguard to prevent from being diverted and directly fashioned into bombs. This number is not only growing, but at a rate far faster than that of the IAEA's safeguards budget. The amount of separated plutonium and highly enriched uranium (nuclear fuels that can be fashioned into bombs in a matter of hours or days) that the IAEA inspects, for example, has grown more than six-fold between 1984 and 2004 while the agency's safeguards budget has barely doubled (see the Figure 1 below).

Meanwhile, the number of nuclear fuel fabrication and fuel making plants (facilities that are by far the easiest to divert nuclear material from) has grown in the last 2 decades from a mere handful to 65. Then, there is the number of other plants containing special nuclear material that the IAEA must safeguard: It has roughly tripled to more than 900 facilities today.[6]

These trends have forced the IAEA to work their inspections staff much harder. Over the last 20 years,

- From 1984 to 2004, IAEA safeguards spending roughly $105 m in constant '04 dollars.

- Amounts of HEU and separated Pu, meanwhile, grew nearly 6-fold -- enough to make 12,000 to 21,000 crude nuclear weapons

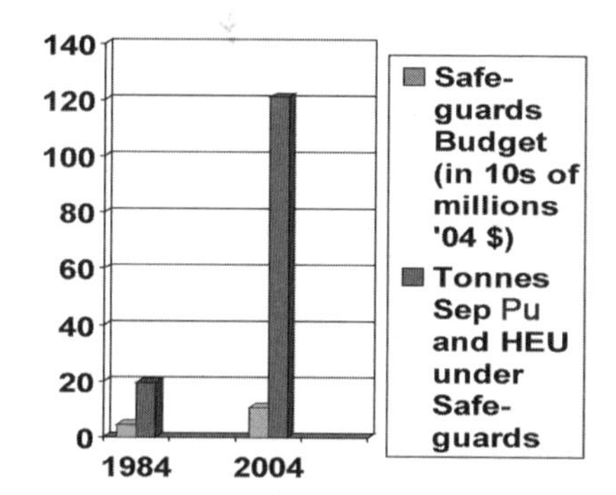

HEU: Highly enriched uranium
Pu: Plutonium

Figure 1. IAEA Safeguards Spending vs. Mounting Weapons Usable Material Stockpiles.

the number of days IAEA inspectors have been in the field has nearly doubled from 60 to 70 days to 125 to 150.[7] This doubling has not only cost more money, it is one of the reasons (along with unreasonable employment and contracting rules) for a hollowing out of IAEA's experienced inspections staff. This hollowing out is expected to become acute. As noted by the U.S. GAO, about 50 percent or 30 out of 75 of the IAEA's senior safeguards staff are expected to retire by 2011.[8]

One way to address this inspections crunch is to have the IAEA simply inspect less. This could be done legally by implementing the Additional Protocol. In fact, limiting the number of routine safeguards inspections is one of the incentives the IAEA currently offers countries to sign up to the Additional Protocol. Once a country has ratified the Additional Protocol and the IAEA has established that there "is no indication of undeclared nuclear material activities for the state as a whole," the agency can reduce the number of routine nuclear inspections it makes of that country's nuclear materials and facilities significantly.[9]

The trouble with taking this approach, though, is that initially it actually *increases* the amount of staff time and resources that the IAEA would have to spend to safeguard a given country. It turns out that determining whether or not a country has no undeclared nuclear materials activities takes considerable safeguards staff resources.[10] Over the entire lifetime of a nuclear facility (i.e., 20 to 50 years), then, applying integrated safeguards might reduce the total amount of staff time needed to safeguard a particular set of nuclear plants slightly but in the first few years, more, not less staff time and safeguards resources would be consumed.[11]

Also, the Additional Protocol authorizes the IAEA to conduct wide area surveillance inspections. These would be extremely useful in the case of Iran or North Korea. They also would require significant additional safeguards staff and funding (by one estimate done for NPEC by a seasoned former IAEA inspector, perhaps a plus up in funding constituting as much as 30 percent of the IAEA's entire current safeguards budget).[12] So far, the IAEA has done nothing to establish such an inspections capability.

Finally, relying heavily on integrated safeguards may be unsound in principle. As already noted, they require the IAEA to determine that the country in question has no undeclared nuclear material. Yet, the IAEA's safeguards staff itself has admitted that it cannot yet be relied upon to discover covert nuclear fuel making facilities in the hardest cases (e.g., Iran). Also, reducing the frequency of on-site inspections increases the risks that a member state might divert materials to make bombs without the IAEA finding out until it is too late.

In a detailed study completed for NPEC late in 2004 on the proliferation risks associated with light water reactors, several scenarios were presented under which fresh and spent nuclear fuel rods might be diverted to make nuclear weapons fuel in covert reprocessing or enrichment plants in a matter of days or weeks without tipping off IAEA inspectors.[13] These scenarios were subsequently validated independently by key officials working within the IAEA's Standing Advisory Group on Safeguards, the U.S. Department of State, Los Alamos National Laboratory.[14]

That a country could evade IAEA inspectors in diverting entire fuel rods is disquieting. One would assume that the current crop of IAEA remote nuclear monitoring equipment could be counted upon entirely to warn against such diversions. In fact, they cannot.[15] Most of the currently deployed remote sensors do not allow the IAEA even to know day to day if these systems are on. This is a serious shortcoming. Over the last 6 years, the agency has learned of camera "blackouts" that lasted for "more than 30 hours" on 12 separate occasions. What is worse, it only learned of these blackouts *after* inspectors went to the sites and downloaded the camera recordings as they are required to do every 90 days.[16]

Under new proposed "integrated safeguards" procedures, such "downloading," moreover, would occur as infrequently as every 12 months—a period within which a state could conceivably make a nuclear weapon unbeknownst to the IAEA.[17] The IAEA staff recently proposed to correct this inspections gap by accelerating implementation of near real-time monitoring using satellite communication connections. This effort, though, is still being implemented at an excruciatingly slow pace due to a lack of funds.[18]

STRUCTURAL PROBLEMS

The current gap in the IAEA's near-real time monitoring capabilities may be worrisome but it, at least, can be addressed assuming additional safeguards funding is made available. Far more intractable is the IAEA's inability to detect diversions in a timely manner from nuclear fuel making plants. As already noted, NPEC's earlier study on the proliferation dangers associated with light water reactors highlighted the relative ease with which states might build covert reprocessing plants or divert fresh civilian fuel to accelerate undeclared uranium enrichment efforts.

Additional NPEC-commissioned research detailed just how poorly IAEA safeguards have performed at nuclear fuel plants in Europe and Japan. In his study, "Can Nuclear Fuel Production in Iran and Elsewhere Be Safeguarded against Diversion,"[19] Dr. Edwin Lyman highlights several examples. At a fuel fabrication plant at Tokai-mura in Japan making mixed-oxide (MOX) fuel out of powdered uranium and nuclear weapons usable separated plutonium, the IAEA could not account for 69 kilograms of plutonium. This is enough to make at least nine nuclear weapons (assuming the IAEA's eight kilograms per weapon estimate) or twice that figure (assuming the U.S. Department of Energy's more accurate four kilograms per crude nuclear weapon figure). Only after 2 years, the expenditure of $100 million, and the disassembling of the plant could the operator claim that he could account for all but 10 kilograms (i.e., one to two bombs' worth).[20]

Dr. Lyman details a similarly disturbing incident involving MOX scrap in Japan where at least one bomb's worth of weapons-usable plutonium went missing and another accounting discrepancy at a Japanese reprocessing plant at which the IAEA lost track of between 59 and 206 kilograms of bomb-usable plutonium (but only was able to determine this years *after* the material initially went unaccounted for). Add to these discoveries the many bombs' worth of material unaccounted for (MUF) annually at reprocessing plants in France and the United Kingdom (where the IAEA has employed its very latest near-real time monitoring techniques), and there's cause for alarm.[21]

The picture relating to safeguarding centrifuge enrichment plants is not much brighter. Even at plants where IAEA monitoring and inspectors are on site, there will be times in between inspections during which remote monitoring might be defeated. There also is the constant problem of the operator giving false design, production, or capacity figures.[22]

In any case, the times between a decision to divert and having enough material to make a crude bomb (assuming the IAEA's high estimate of 25 kilograms of highly enriched uranium being required to make one weapon) are so short, even an immediate detection of the diversion, which is by no means assured, would generally come too late to afford enough time to prevent bombs from being made. In the case of a small commercial sized plant, a bomb's worth could be made in as little as 18 hours to 12 days (depending on whether natural or slightly enriched uranium is used as feed).[23]

SAFEGUARDS ASSUMPTIONS

Exacerbating this safeguards gap is the IAEA's overly generous view of how much material must be diverted to make a bomb (referred to by the IAEA as a "significant quantity") and how long it might take to convert this material into a nuclear weapon (known as the "conversion time"). Most of these IAEA estimates were made over 30 years ago. To reassess their accuracy, NPEC commissioned Thomas Cochran, chief nuclear scientist at the Natural Resources Defense Council (NRDC). His analysis and conclusions were revealing. The IAEA estimates it would take eight kilograms of separated plutonium and 25 kilograms of highly enriched uranium to make a crude bomb. These estimates were found to be too high by a minimum of 25 percent and a maximum of 800 percent, depending on the weapons expertise employed and the yield desired (see Figure 2 below).[24]

	Weapon-Grade Plutonium (kg)			Highly-Enriched Uranium (kg)		
Yield (kt)	Technical Capability			Technical Capability		
	Low	Medium	High	Low	Medium	High
1	3	1.5	1	8	4	2.5
5	4	2.5	1.5	11	6	3.5
10	5	3	2	13	7	4
20	6	3.5	3	16	9	5

Values rounded to the nearest 0.5 kilogram.

Figure 2. NRDC Estimate of the Approximate Fissile Material Requirements for Pure Fission Nuclear Weapons.

When presented with these figures, senior IAEA safeguards staff did not dispute them. Instead they argued that the "exact" amount of diverted nuclear material needed to make a crude bomb was not that important. Instead, what mattered most was the IAEA's ability to detect microscopic amounts of weapons-usable materials since securing such environmental samples was the thing most likely to put an inspected party in the international spotlight.[25]

The potential downside of taking this approach, however, is significant. It is these estimates, along with the agency's projections of how long it takes a proliferator to convert uranium and plutonium materials into bombs (i.e., conversion times), that the IAEA uses to determine how often it should conduct its inspections of different nuclear facilities. If these estimates are too high, the frequency of inspections needed to detect military diversions risks is egregiously low. Certainly, what the IAEA defines as desirable "detection times"—the maximum time that may elapse between the diversion of a significant quantity of nuclear material and the likely detection of that diversion—should correspond (according to the IAEA's own guidelines) to the agency's estimated conversion times. If they don't, IAEA-inspected countries could count on being able divert a crude weapon's worth of nuclear material and fashioning it into a bomb before the IAEA could either detect the diversion or have any chance of taking appropriate action to block bomb making.

This worry seems quite real when one considers how high the IAEA's 30-year old significant quantity estimates appear to be and one then looks at how generous the IAEA's estimated conversion times are (see Figure 3 below).

Using the history of the Manhattan Project as a benchmark, the IAEA's first set of estimates regarding the amount of time (7 to 10 days) needed to convert separated plutonium (Pu) or highly enriched uranium (HEU) or ^{233}U metal were judged by Dr. Cochran to be the correct order of time. The key reason why is that in 1945, the plutonium and enriched uranium for the first American bombs had to be shipped thousands of miles from where they were produced to where

Beginning Material Form	Conversion Time
Pu, HEU, or ^{233}U metal	Order of days (7-10)
PuO_2, $Pu(NO_3)_4$ or other pure Pu compounds: HEU or ^{233}U oxide or other pure U compounds; MOX or other nonirradiated pure mixtures containing Pu, U ($^{233}U + ^{235}U > 20$ percent); Pu, Heu, and/or ^{233}U in scrap or other miscellaneous impure compounds	Order of weeks (1-3)*
Pu, HEU, or ^{233}U in irradiated fuel	Order of months (1-3)
U containing <20 percent ^{235}U and ^{233}U; Th	Order of months (3-12)

*This range is not determined by any single factor, but the pure Pu and U compounds will tend to be at the lower end of the range and the mixtures and scrap at the higher end.

Figure 3. Estimated Material Conversion Times for Finished Pu or U Metal Weapons Components.[26]

the material was fashioned into nuclear weapons. This transport took several days. If a country making nuclear weapons did not have to ship these distances, the conversion time could be much shorter. However, the conversion times could still be on the order of a day or more.

The IAEA's estimates of how long it would take (1 to 3 weeks) to convert fresh plutonium-uranium fuels (known as mixed oxide fuels or MOX) do not fare as well. Here, Dr. Cochran points out that it would take no more than a week and possibly as little as a few days to convert these materials into metal bomb components. Instead of a matter of weeks, he concludes that the correct conversion time should be measured in a matter of days.

As for the IAEA's conversion time estimates of 1 to 3 months for plutonium, HEU, or ^{233}U contained in irradiated spent reactor fuel, these were also judged to be accurate *only* if the country possessing these materials did not have a covert or declared reprocessing or enrichment plant. If the country in question did, then it could possibly convert the spent fuel into bombs in a matter of weeks rather than months.

Finally, Dr. Cochran agreed with the IAEA's low end estimated conversion time of 3 months for low enriched uranium but, with the increased international availability of gas centrifuge uranium enrichment technology, found the IAEA's high end estimate of 12 months to be totally unwarranted. In fact, as already noted, a country might well be able to convert low enriched uranium into a bomb in a matter of weeks or less.[27]

The policy ramifications of these overly generous IAEA estimates are significant. They directly impact what the IAEA's detection goals should be. In three cases—the conversion of low enriched uranium; the conversion of plutonium, HEU, and ^{233}U metal; and of these materials in spent fuel—the order of time associated with the IAEA estimates is correct. In another three cases, however—the conversion of plutonium, HEU and ^{233}U in MOX; and of these materials in spent fuel; and of low enriched uranium if the inspected country has covert or declared nuclear fuel making facilities—the IAEA's estimates are egregiously high. IAEA conversion times are measured in months when they should be measured in weeks, and in weeks when they should be measured in days.

As a result, the IAEA's timeliness detection goals in many cases are dangerously high. More important, the agency's current detection goals give the mistaken impression that the IAEA can detect military diversions before they result in bombs or even early enough to prevent the diversion from succeeding when this clearly is not the case. Dr. Cochran's analysis highlights that timely detection for plutonium, HEU, and ^{233}U in metal and in fresh MOX is simply not possible. He concludes that countries that do not yet have nuclear weapons should not be allowed to stockpile or produce these materials. He reaches the same conclusion regarding the agency's ability to detect diversions of plutonium, HEU, and ^{233}U in nonweapons states that may have a declared or covert enrichment or reprocessing plant. In these cases, the problem is not that the IAEA's timeliness detection goals are too liberal; it is that the IAEA claims that timely detection is possible at all (see Figure 4 below).

To some extent, these critical conclusions are gaining official support. As the IAEA's former director for safeguards recently explained, when it comes to nuclear fuel making, the IAEA is must rely on its limited ability to ascertain the inspected country's military intent. [28] Even the director general of the IAEA conceded that once a country acquires separated plutonium and HEU, the IAEA must rely on these states' continued peaceful intentions, which could change rapidly. Unfortunately, the IAEA's Board of Governors and major governments, including the United States, do not yet fully appreciate the full implications of these points.

If the IAEA cannot provide timely detection of diversions of weapons-usable HEU and plutonium from centrifuge enrichment, spent fuel reprocessing, and other fuel-making plants, how can it claim that it is "safeguarding" such facilities in Brazil, the Netherlands, Germany, and Japan? How can it

MATERIAL	IAEA Conversion Time	Cochran/ NPEC Commissioned Estimate	Official IAEA Timeliness Detection Goal	NPEC Conclusions and Recommended Timeliness Detection Goals
Pu, HEU, ^{233}U in metal form	Order of days (7-10)	Order of days (7-10)	1 month	Timely detection is not possible
In fresh MOX	Order of weeks (1-3)	Order of days (7-10)	1 month	Timely detection is not possible
In irradiated spent fuel	Order of months (1-3)	Order of months (1-3), if reprocessing - enrichment plant on tap (7-10 days	3 months	For countries with covert or declared nuclear fuel making plants, timely detection is not possible
Low enriched uranium	Order of months (3-12)	Order of weeks to months	1 year	For countries with covert or declared enrichment plants, timely detection is not possible

Figure 4. IAEA's Timeliness Detection Goals and NPEC's Conclusions.

effectively safeguard an Indian reprocessing plant (as is being currently proposed by the Indian government as a way to allow for the reprocessing of foreign fuel for use in an unsafeguarded Indian breeder reactor)? What of the idea of promoting regional nuclear fuel-making centers in nonweapons states, such as Kazakhstan? How might the IAEA prevent diversions?[29]

What of other more ambitious missions for the IAEA? If one cannot keep track of many bombs' worth of nuclear weapons-usable material produced annually at declared civilian nuclear fuel-making plants or assure that the plants themselves would not be seized, how much sense does it make to encourage the IAEA to oversee an even more difficult to verify military fissile production cut-off treaty?[30] Finally, there is the question of large research reactors and nuclear power plants, which require lightly enriched fuel or produce significant quantities of plutonium. If the IAEA cannot reliably ferret out covert nuclear fuel making programs, how safe is it to export such machines to new countries, particularly in war-torn regions, such as the Middle East?

The questions here are all intentionally rhetorical. Yet, many experts and officials within the IAEA and the U.S. and other governments actively support at least one or more of the questionable nuclear initiatives referred to. This needs to change.

One of this report's key recommendations is to encourage governments and the IAEA to reassess the agency's estimates of what a significant quantity is, along with the conversion times for various materials and what the proper detection goals should be for the agency. The most important part of this reassessment would be to clarify precisely what nuclear material diversions the agency cannot be counted upon to detect in a timely fashion. At a minimum, this should include the possible diversion of HEU, ^{233}U, and MOX from storage facilities, reprocessing plants, enrichment plants, fuel fabrication plants and of direct-use materials from large research or power reactors in nonweapons states that might have covert or declared nuclear fuel-making plants.

For these nuclear activities and materials, the IAEA would do well simply to declare that the agency can monitor, but not safeguard them—i.e., that it can mind these facilities and materials but not assure detection of their possible military diversion in a timely fashion. Such an honest announcement would be helpful. First, it would put governments on notice about how dangerous the conduct of certain nuclear activities most closely related to bomb making actually are. Second, it would encourage countries to demand more monitoring and physical security of these unsafeguardable nuclear materials and activities. The primary aim in increasing such security and monitoring would not be to block diversions so much as to increase the chance of at least detecting them after they had occurred. This would help to deter such deeds and to limit further the risks of nuclear theft or sabotage. It is difficult to determine what the optimal level of monitoring and physical security might be for this purpose. But a good place to start would be to upgrade physical security at nuclear facilities that handle or produce nuclear weapons-usable materials to those security standards currently employed at the most secure nuclear weapons production and storage facilities.

FUNDING

As already noted, the IAEA's inspections of safeguardable nuclear materials and activities could be enhanced in a number of ways. More near-real time monitoring could significantly enhance the agency's ability to detect the diversion of fuel rods. Retention and increasing the numbers of experienced nuclear inspectors could help assure the IAEA actually meets its timelines detection goals and is able to analyze remote sensing information and imagery properly. Full support for the IAEA's environmental sampling activities would enable it to replace its aging Safeguards Analytical Laboratory and help the IAEA shorten the time needed to analyze samples from months to days or weeks. Much needed work to develop new safeguarding research capabilities and equipment could proceed much more quickly if more funds were made available.[31] Similarly, with proper funding, the IAEA could muster reserve inspections staff and resources to meet unexpected demands and to provide the agency with deployable wide-area surveillance capabilities.

The first step to address these current gaps is simply to admit that they exist. For years, the IAEA has avoided doing this publicly. At the very outset of NPEC's investigations, early in 2005, the IAEA's safeguards planning staff briefed NPEC that it believed safeguards funding for the mid-term (i.e., the next 5 years) was sufficient. It conceded that it had given little or no thought to what funding agency safeguards might require beyond this period.

Fortunately, in the last 2 years, the agency's approach to safeguards planning has improved. Most recently the IAEA's director general highlighted the agency's lack of safeguards funding to deal with urgent inspections requirements associated with monitoring the shutdown of the reactor in North Korea. In a statement he made on July 9, 2007, IAEA Director General Dr. Mohamed ElBaradei explained that the IAEA was having difficulty paying for the nearly 4 million euros needed to cover the monitoring costs. He went on to note:

> The DPRK case clearly illustrates the need for the agency to have an adequate reserve that can be drawn upon to enable it to respond promptly and effectively to unexpected crises or extraordinary requests, whether in the areas of verification, nuclear and radiological

> accidents, or other emergencies. The agency's financial vulnerability is also demonstrated by our current cash situation, which indicates that unless some major donors pay their outstanding contributions by the end of next month, the agency will have to draw from the Working Capital Fund in order to continue operations. And unless contributions are received by September, that Fund would be depleted. Finally, let me stress that the recent process of preparing and getting approval for the programme and budget for the next biennium has once again highlighted the urgent need for adequate resources to ensure effective delivery of the entire programme that you have requested. As I made clear during the last Board, even with the budget originally proposed by the Secretariat, the agency remains under-funded in many critical areas, a situation which, if it remains unaddressed, will lead to a steady erosion of our ability to perform key functions, including in the verification and safety fields.

At the conclusion of this statement, the director general then announced that he had initiated a study to examine the IAEA's "programmatic and budgetary requirements" over the "next decade or so." In addition, he announced his intention to create a high level panel to study options for financing the agency's requirements.[32]

The director general's announcement accords almost precisely to the recommendations Dr. Thomas E. Shea made to a select group of U.S. and European officials, including Dr. ElBaradei's top scientific advisor, Andrew Graham, at an NPEC-sponsored conference held in Paris, France, on November 13, 2006.[33] In his brief, "Financing IAEA Verification of the NPT," Dr. Shea argued that North Korea "provides a clear justification" for additional safeguards funding and that to secure it the director general "should convene a council of wise men to assist in determining how best to respond in this matter."

As has been noted, the IAEA's funding is based on a United Nations formula that weights a country's gross domestic product and other factors. This formula may be sensible for raising general funds, but for nuclear safeguards purposes it produces several anomalies. Countries with no large reactors (e.g., Italy) are sometimes asked to pay in more than countries that have a score or more of them (e.g., the Republic of Korea). The UN assessment method also overlooks the actual inspections requirements particular nuclear facilities impose that are significantly higher than the norm. Nuclear fuel-making plants of any type, reactors that are on-line fueled (i.e., fueled constantly while they are operating, e.g., heavy water and gas-cooled reactors, versus off-line fueled reactors, e.g., light water reactors), and fast reactors all impose additional inspections challenges that are significantly higher than other types of nuclear facilities. Inspecting or monitoring these facilities costs much more than it does for other nuclear plants, yet the operator or owner pays no premium to cover these additional expenses.

Finally, because the IAEA's current approach to assessing its members for contributions fails to raise enough money for the Department of Safeguards, the agency must depend on additional voluntary contributions of cash and technical assistance. Almost all of voluntary contributions come from the United States (amounting to roughly 35 percent of the IAEA's safeguards budget). That so much of the safeguards budget is paid for voluntarily by the United States is politically awkward, since the agency's most challenging inspections cases—e.g., India, Iran, North Korea, Taiwan, and South Korea—are all of special interest to Washington.[34]

Dr. Shea suggests several ways to increase funding for safeguards—from setting up an endowment to selling bonds. All of them are worth pursuing, but one of his ideas is particularly deserving: the customer (i.e., the inspected party) should pay. There already is a precedent for doing this. Taiwan, which the IAEA does not recognize as being an independent, sovereign nation does not pay as other nations do but instead pays what the IAEA estimates it costs the agency to inspect Taiwan's plants.

This report recommends that the United States take the lead getting the IAEA to help fund its safeguards activities with a user fee. The United States should continue to make its voluntary contributions but instead of making them as it currently does, Washington should justify them as representing a specific percentage of costs associated with generating nuclear electricity annually in the United States. Japan, which also gives voluntary contributions, should be urged to do likewise. Agreement might subsequently be reached on an international standard and this surcharge should be tacked on to the cost of electricity or other products these civilian plants produce. The last step would be to make the surcharge obligatory and assign all of the funds so raised to the IAEA's Department of Safeguards.

In addition to these funds, the agency should consider assessing an additional charge for the monitoring of unsafeguardable nuclear materials or facilities (e.g., nuclear fuel-making plants and nuclear weapons or near-nuclear weapons-usable fuels, etc.). Finally, an additional fee might be levied against nuclear facilities or plants that are particularly costly for the IAEA to meet its own timeliness detection goals (e.g., for on-line fueled reactors).

RIGHTS

Some countries, of course, are likely to bridle at these proposals, arguing that imposing surcharges would interfere with their right to peaceful nuclear energy. These arguments, however, should be rejected. The exercise of one's right to develop, research and produce peaceful nuclear energy hardly extends to not paying what it costs to safeguard these activities against military diversion. Also, the premise behind these arguments is a dangerously distorted view of the nuclear rules—that so long as states can claim a nuclear material or activity has some conceivable civilian application, any country has a right to acquire or engage in them even if they are unprofitable commercially, bring their possessor to the very brink of having bombs, and cannot be safeguarded against military diversion. The danger of this over-generous interpretation of the NPT is obvious: It risks, as U.N. General Secretary Koffi Anan explained to the 2005 NPT review conference, creating a dangerous world full of nuclear fuel-producing states that claim to be on the right side of the NPT, but are, in fact, only months or even days from acquiring nuclear weapons.[35]

Luckily, as research conducted for NPEC makes clear, this interpretation of the NPT is wrong.[36] The NPT makes no mention of nuclear fuel making, reprocessing, or enrichment. Spain, Romania, Brazil, and Mexico all tried in the late 1960s to get NPT negotiators to make it a duty under Article IV for all of the nuclear supplier states to supply "the entire fuel cycle" including fuel making, to nonweapons states. Each of their proposals was turned down.[37] At the time, the Swedish representative to the NPT negotiations even suggested that rules needed to be established to *prevent* nations from getting into such dangerous activities, since there seemed no clear way to prevent nations that might make nuclear fuel from quickly diverting either the fuel or the fuel making plants very quickly to make bombs.[38] They certainly were not interested in protecting uneconomical propositions that are unnecessary and that could bring states to the brink of having bombs.[39]

A clear case in point was the NPT's handling of peaceful nuclear explosives, which turned out to be so dangerous and impossible to safeguard that the treaty spoke only of sharing the "potential benefits" of peaceful nuclear explosives that would be supplied by nuclear weapons states. No effort, however, was ever made to request or to offer such nuclear explosives because they were so costly to use as compared to conventional explosives and no clear economic benefit could be found in using them.[40]

Finally, in no case did the framers of the NPT believe that the inalienable right to develop, research or produce peaceful nuclear energy should allow states to contravene the NPT restrictions designed to prevent the proliferation of nuclear weapons. These restrictions are contained in articles I, II, and III of the treaty. Article I prohibits nuclear weapons states "assist[ing], encourage[ing], or induc[ing] any non-weapons state "to manufacture or otherwise acquire" nuclear weapons. Article II prohibits non-weapons states from acquiring in any way nuclear explosives or seeking "any assistance" in their manufacture. Together these two prohibitions suggest that the NPT not only bans the transfer of actual nuclear explosives, but of any nuclear technology or materials that could "assist, encourage or induce" nonweapons states to "manufacture or otherwise acquire" them.[41]

If there was any doubt on this point, the NPT also requires all nonweapons states to apply safeguards against all of their nuclear facilities and holdings of special nuclear materials. The purpose of these nuclear inspections, according to the treaty is "verification of the fulfillment of its obligations assumed under this Treaty with a view to preventing diversion of nuclear energy from peaceful uses to nuclear weapons."[42] It was hoped at the time of the treaty's drafting that a way could be found to assure such safeguards. It, however, was not assumed that such techniques already existed.[43]

CONCLUSION

It would be useful to remind members of the IAEA of these points. The most direct and easiest way to begin is to make clear what can and cannot be safeguarded—i.e., what can and cannot be monitored so as to detect a military diversion *before* it is completed. Beyond this, the IAEA should have the owner, operators, and customers of nuclear facilities bear the costs associated with monitoring and safeguarding them. The hope here would be that the poor economics associated with large nuclear power reactors and nuclear fuel making plants might help some nations reconsider the desirability of acquiring them. Making sure that the full external costs of IAEA inspections are carried by each inspected party would be useful. The NPT, after all, is dedicated to sharing the "benefits" of peaceful nuclear energy, not money losing programs that bring countries to the brink of having bombs.[44]

In this regard, it is worth noting that a popular idea to promote nonproliferation that enjoys IAEA support—assuring supplies of nuclear fuel at "affordable" or "reasonable" prices with fuel banks and the construction of fuel making centers in nonweapons states—could, under certain circumstances, actually undermine the NPT's intent. If these assurances come with subsidies, more countries may be enticed to develop large nuclear programs that may not be economically viable. If these assurances come, as they now do, with repeated pledges that the recipients of the fuel retain a per se right to make nuclear fuel any time they wish, then, there also is a danger that after bootstrapping themselves up with fuel assistance, recipient nations will simply proceed to make fuel on their own. Finally, if the assurances result in building fuel-making centers in countries that do not yet have nuclear weapons, the risks of nuclear weapons proliferation will surely increase.[45]

Unfortunately, there is no technical fix yet for the dangers associated with declared and covert nuclear fuel making activities. Initially, one of the claims of the U.S. Global Nuclear Energy Partnership (GNEP) initiative was that it would make it possible to recycle spent fuel in a proliferation resistant manner and, thereby, strengthen the international nonproliferation regime. NPEC commissioned two leading national nuclear experts at MIT and Princeton to examine these claims.[46] Their conclusion—that these assertions do not hold up and that the recycling technology would be more not less difficult to monitor—now is closer to the view that even the Department of Energy itself is making. Its official strategy document now warns against spreading its "proliferation resistant" uranium extraction (UREX) system for fear it, too, might be diverted to make bombs.[47]

Finally, routine inspections alone are unlikely to deter states from breaking the rules. One of the key reason why is that after the agency's experience with Iraq, Iran, and North Korea, it is no longer clear what might happen to the next nation that breaks its IAEA safeguards agreement or the strictures of the NPT. Pierre Goldschmidt, the former IAEA Deputy Director who headed up the agency's Department of Safeguards, knows this first hand: He had to deal with Iraq, Iran, and North Korea where the burden of proof for misbehavior was laid at IAEA's doorstep rather than with the suspect party. NPEC was fortunate to be able to commission Dr. Goldschmidt to review what might be done to correct this. His recommendations, which consist of developing a set of country-neutral rules that come into play when the IAEA is *unable* to clarify suspicious behavior or when a majority of the IAEA board finds a nation to be non-compliant or attempting to break free from the NPT before it is found to be in compliance, are among the ones contained in this final report.

Adoption of these recommendations, along with the others, is essential to give the IAEA the resources and authorities it needs to succeed. Beyond this, member states must stop pushing the IAEA to safeguard nuclear materials and projects that are both unnecessary and so close to bomb making that no agency, national or international, could credibly safeguard them against military misuse. The balance, in short, that must be struck is to give the agency much more to do its job and to back off demanding that it tackle the impractical.

ENDNOTES - CHAPTER 1

1. For the latest of these, see U.S. Government Accountability Office, *Nuclear Nonproliferation: IAEA Has Strengthened Its Safeguards and Nuclear Security Programs, but Weaknesses Need to Be Addressed*, GAO-06-93, Washington, DC, October 7, 2005, available from *www.gao.gov/new.items/d0693.pdf*.

2. Upon the suggestion of the United States, the IAEA Board of Governors agreed in 2005 to create a special committee to advise how best to strengthen the IAEA's current nuclear safeguards system. In private interviews, NPEC has learned that the confidential recommendations of this committee were almost entirely administrative and quite modest in scope. None of the recommendations, which were made in the spring of 2007, have yet been acted upon.

3. These states included Algeria, Tunisia, Morocco, Libya, Egypt, Turkey, Jordan, Saudi Arabia, Yemen, Vietnam, Australia, Indonesia, Bangladesh, Vietnam, and Nigeria.

4. For a list of experts who participated in NPEC's IAEA safeguards workshops, go to Appendix I.

5. This line of argument was actually presented to NPEC's executive director in a private briefing by the IAEA safeguards planning staff in Vienna early in 2006.

6. For data on the IAEA's safeguards budget obligation in current—not constant—U.S. dollars, see *The Agency's Accounts for 1984*, GC(XXIX)/749, p. 26; and *The Agency's Accounts for 2004*, GC(49)/7, p. 47. For data on the amount of nuclear material safeguarded by the IAEA, see *Annual Report for 1984*, GC(XXIX)/748 (Vienna, Austria: IAEA, July 1985), p. 63; and *Annual Report for 2004*, GC(49)/5, Annex, Table A19.

7. Private interviews with safeguards staff and former IAEA safeguards inspectors at the Los Alamos National Laboratory, Los Alamos, NM, May 12, 2005.

8. See Gene Aloise, Director Natural Resources and Environment, U.S. Government Accountability Office, "Nuclear Nonproliferation: IAEA Safeguards and other Measures to Halt the Spread of Nuclear Weapons Materials," testimony before the Subcommittee on National Security, Emerging Threats and International Relations, Committee on Government Reform, House of Representatives, September 26, 2006.

9. For a more detailed discussion of the Additional Protocol see Richard Hooper, "The IAEA's Additional Protocol," *Disarmament Forum*, "On-site Inspections: Common Problems, Different Solutions," 1999, No. 3, pp. 7-16, available from *www.unidir.ch/bdd/fiche-article.php?ref_article=209*.

10. For example, in the case of Japan, the IAEA needed five years to determine it had no undeclared nuclear material activities and estimates that it will need about as much time to make the same determination for Canada. See U.S. GAO, *Nuclear Nonproliferation*, pp. 12-13.

11. According to IAEA internal analyses, the average lifetime savings in safeguards resources likely implementing integrated safeguards may be no more than five percent. See C. Xerri and H. Nackeaerts on behalf of the ESARDA Integrated Safeguards Working Group, "Integrated Safeguards: A Case to Go Beyond the Limits: Consequences of Boundary Limits Set to the Reduction of "Classical Safeguards Measures on Efficiency and Resources Allocation in Integrated Safeguards" produced in 2003 for the IAEA, available from *esarda2.jrc.it/bulletin/bulletin_32/06.pdf*. For an official overview of the various safeguards resources required to implement the Additional Protocol, see Jill N. Cooley, "Current Safeguards Challenges from the IAEA View," an IAEA document produced in 2003, available from *esarda2.jrc.it/events/other_meetings/inmm/2003-esarda-inmm-Como/1-paper%20pdf/1-1-040127-cooley.pdf*.

12. See Garry Dillon, "Wide Area Environmental Sample in Iran," available from *www.npec-web.org/Essays/WideAreaEnvironmentalSampling.pdf*.

13. See Victor Gilinsky, Harmon Hubbard, and Marvin Miller, *A Fresh Examination of the Proliferation Dangers of Light Water Reactors*, Washington, DC: The Nonproliferation Policy Education Center, October 22, 2004, reprinted in Henry Sokolski, ed., *Taming the Next Set of Strategic Weapons Threats*, Carlisle, PA: Strategic Studies Institute, U.S. Army War College, 2006, available from *www.npec-web.org/Frameset.asp?PageType=Single&PDFFile=20041022-GilinskyEtAl-LWR&PDFFolder=Essays*.

14. See, e.g., Andrew Leask, Russell Leslie and John Carlson, "Safeguards As a Design Criteria—Guidance for Regulators," Canberra, Australia: Australian Safeguards and Non-proliferation Office, September 2004, pp. 4-9 available from *www.asno.dfat.gov.au/publications/safeguards_design_criteria.pdf.*

15. For more detailed discussion of how fuel diverted from different commercial and research reactors could help accelerate a country's covert bomb program, go to Appendix II of this report.

16. See J. Whichello, J. Regula, K. Tolk, and M. Hug, "A Secure Global Communications Network for IAEA Safeguards and IEC Applications," IAEA User Requirements Document, May 6, 2005.

17. The problem of states "losing" fuel rods, it should be noted is not limited to countries intent on diverting them to make bombs. The U.S. civilian nuclear industry, which has a clear industrial interest in keeping track of its nuclear fuel, has had difficulty keeping proper account of all of it. On this point, see U.S. Government Accountability Office, "NRC Needs to Do More to Ensure Power Plants Are Effectively Controlling Spent Fuel," GAO O5-339, April 2005, available from *www.gao.gov/new.items/d05339.pdf.*

18. Only about a third of the facilities at which the IAEA currently has remote sensors have near-real-time connectivity with Vienna or other regional headquarters and almost all of these facilities are in countries that are of minimal proliferation risk. This information was presented at a private IAEA Department of Safeguards briefing of NPEC's executive director at Vienna, Austria, IAEA Headquarters, January 30, 2006.

19. See Edwin Lyman, "Can Nuclear Fuel Production in Iran and Elsewhere be Safeguarded Against Diversion," presented at NPEC's Conference "After Iran: Safeguarding Peaceful Nuclear Energy" held in London, UK, October 2005, available from *www.npec-web.org/Frameset.asp?PageType=Single&PDFFile=Paper050928LymanFuelSafeguardDiv&PDFFolder=Essays.*

20. This incident received only scant public attention. See, however, Bayan Rahman, "Japan 'Loses' 206 kg of Plutonium," *Financial Times*, January 28, 2003, available from *news.ft.com;servlet/*

ContentServer?pagename=FT.com/StoryFT/FullStory&c=StoryFT&cid=1042491288304&p=10112571727095.

21. *Idem.* Also see "Missing Plutonium 'Just on Paper,'" *BBC News*, February 17, 2005, available from *news.bbc.co.uk/1/hi/uk/4272691.stm*; and Kenji Hall, "Missing Plutonium Probe Latest Flap for Japan's Beleaguered Nuclear Power Industry," *Associated Press* (Tokyo), January 28, 2003, available from *www.wise-paris.org/index.html?/english/othersnews/year_2003/othersnews030128b.html&/english/frame/menu.html&/english/frame/band.html.*

22. These points have been long recognized by outside experts. See Paul Leventhal, "Safeguards Shortcomings—A Critique," Washington, DC: NCI, September 12, 1994; Marvin Miller, "Are IAEA Safeguards in Plutonium Bulk-Handling Facilities Effective?" Washington, DC: NCI, August 1990; Brian G. Chow and Kenneth A. Solomon, *Limiting the Spread of Weapons-Usable Fissile Materials*, Santa Monica, CA: RAND, 1993, pp. 1-4; and Marvin Miller, "The Gas Centrifuge and Nuclear Proliferation," in *A Fresh Examination*, p. 38.

23. See *idem.* and the comments of the former chairman of the IAEA's Standing Advisory Group on International Safeguards, John Carlson, Australian Safeguards and Non-Proliferation Office, "Addressing Proliferation Challenges from the Spread of Uranium Enrichment Capability," Paper prepared for the Annual Meeting of the Institute for Nuclear Materials Management, Tucson, Arizona, July 8-12, 2007. Copy on file at NPEC.

24. See Thomas B. Cochran, "Adequacy of IAEA's Safeguards for Achieving Timely Warning, " paper presented before a conference cosponsored by NPEC and King's College, "After Iran: Safeguarding Peaceful Nuclear Energy," October 2-3, 2005, London, United Kingdom, available from *www.npec-web.org/Frameset.asp?PageType=Single&PDFFile=Paper050930CochranAdequacyofTime&PDFFolder=Essays.*

25. Interview of senior advisors to the IAEA Director General, IAEA Headquarters, Vienna, Austria, January 30, 2006, and January 17, 2005.

26. IAEA, *IAEA Safeguards Glossary*, 2001 Edition, Paragraph 3.13, Figure 3 here is identified as Table I in the IAEA glossary.

27. Cf. the low-end conversion time estimates for low enriched uranium of John Carlson in note 22 above, which for a small commercial enrichment facility range between 18 hours and 12 days.

28. See the testimony of Pierre Goldschmidt before a hearing of the House Subcommittee on National Security and Foreign Affairs of the House Committee on Oversight and Government Reform, "International Perspectives on Strengthening the Nonproliferation Regime," June 26, 2007, Washington, DC, available from the subcommittee upon request.

29. See Ann Mac Lachian, Mark Hibbs, and Elaine Hiruo, "Kazakh Buy-in to Westinghouse Seen as Win-win for Kazakhs, Toshiba," *Nucleonics Week*, July 12, 2007, p. 1; and Kenneth Silverstein, "As North Korea Gives Up Its Nukes, Kazakhstan Seeks a Nuclear Edge," *Harper's Magazine*, July 2007, available from *harpers.org/archive/2007/07/hbc-90000549*.

30. On the challenges of verifying a military fissile production cut-off treaty, see Christopher A. Ford, "The United States and the Fissile Material Cut-off Treaty," delivered *at* the Conference on "Preparing for 2010: Getting the Process Right," Annecy, France, March 17, 2007, available from *www.state.gov/t/isn/rls/other/81950.htm*.

31. For a detailed discussion of what specific new safeguards capabilities the IAEA Department of Safeguards is investigating, see N. Khlebnikov, D. Parise, and J. Whichello, "Novel Technology for the Detection of Undeclared Nuclear Activities," IAEA-CN148/32, presented at the IAEA Conference on Safeguards held in Vienna, Austria, on October 16-20, 2006, available from *www.npec-web.org/Frameset.asp?PageType=Single&PDFFile=20070301-IAEA-NovelTechnologiesProject&PDFFolder=Essays*.

32. See IAEA Director General Dr. Mohamed ElBaradei, "Introductory Statement to the Board of Governors," July 9, 2007, Vienna, Austria, available from *www.globalsecurity.org/wmd/library/news/dprk/2007/dprk-070709-iaea01.htm*.

33. See Thomas E. Shea, "Financing IAEA Verification of the NPT," paper presented before a conference sponsored by the Nonproliferation Policy Education Center and the French Foreign Ministry, "Assessing the IAEA's Ability to Verify the NPT," November 12-13, 2006, Paris, France, available from *www.npec-web.org/Essays/20061113-Shea-FinancingIAEAVerification.pdf.*

34. See U.S. GAO, *Nuclear Nonproliferation,* pp. 34-40.

35. See Statement of UN Secretary General Kofi Annan, Statement to the Nuclear Nonproliferation Treaty Review Conference, May 2, 2005, UN Headquarters, New York, available from *www.acronym.org.uk/docs/0505/doc11.htm.*

36. Robert Zarate, "The NPT, IAEA Safeguards and Peaceful Nuclear Energy: An 'Inalienable Right,' But Precisely To What?" presented at *Assessing the IAEA´s Ability to Safeguard Peaceful Nuclear Energy,* a conference held in Paris, France, on November 12-13, 2006, available from *www.npec-web.org/Frameset.asp?PageType=Single&PDFFile=20070509-Zarate-NPT-IAEA-PeacefulNuclear&PDFFolder=Essays.*

37. See "Mexican Working Paper Submitted to the Eighteen Nation Disarmament Committee: Suggested Additions to Draft Nonproliferation Treaty," ENDC/196, September 19, 1967, in U.S. Arms Control and Disarmament Agency, *Documents on Disarmament, 1967,* Publication No. 46, Washington, DC: U.S. Government Printing Office, July 1968, pp. 394-395; "Romanian Working Paper Submitted to the Eighteen Nation Disarmament Committee: Amendments and Additions to the Draft Nonproliferation Treaty," ENDC/199, October 19, 1967, in *ibid.,* pp. 525-526; "Brazilian Amendments to the Draft Nonproliferation Treaty," ENDC/201, October 31, 1967, in *ibid.,* p. 546; and "Spanish Memorandum to the Co-Chairman of the ENDC," ENDC/210, February 8, 1968, in U.S. Arms Control and Disarmament Agency, *Documents on Disarmament, 1968,* Publication No. 52, Washington, DC: U.S. Government Printing Office, September 1969, pp. 39-40.

38. See "Statement by the Swedish Representative [Alva Myrdal] to the Eighteen Nation Disarmament Committee: Nonproliferation of Nuclear Weapons," ENDC/PV. 243, February 24, 1966, in U.S. Arms Control and Disarmament Agency,

Documents on Disarmament, 1966, Publication No. 43, Washington, DC: U.S. Government Printing Office, September 1967. p. 56.

39. See Eldon V.C. Greenberg, "NPT and Plutonium: Application of NPT Prohibitions to 'Civilian' Nuclear Equipment, Technology and Materials Associated with Reprocessing and Plutonium Use," Nuclear Control Institute, 1984, Revised May 1993.

40. See *Report of Main Committee III*, Treaty on the Nonproliferation of Nuclear Weapons Review and Extension Conference, May 5, 1995, NPT/CONF.1995/MC.III/1, Sec. I, para. 2, emphases added, available from *www.un.org/Depts/ddar/nptconf/162.htm*, which states: "The Conference records that the potential benefits of the peaceful applications of nuclear explosions envisaged in article V of the Treaty have not materialized. In this context, the Conference notes that the potential benefits of the peaceful applications of nuclear explosions have not been demonstrated and that serious concerns have been expressed as to the environmental consequences that could result from the release of radioactivity from such applications and on the risk of possible proliferation of nuclear weapons. Furthermore, no requests for services related to the peaceful applications of nuclear explosions have been received by IAEA since the Treaty entered into force. The Conference further notes that no State party has an active programme for the peaceful application of nuclear explosions."

41. See Greenberg, "NPT and Plutonium"; and Henry D. Sokolski and George Perkovich, "It's Called *Non*proliferation," *Wall Street Journal*, April 29, 2005, p. A16.

42. NPT, Art III, para. 1.

43. For example, see "British Paper Submitted to the Eighteen Nation Disarmament Committee: Technical Possibility of International Control of Fissile Material Production," ENDC/60, August 31, 1962, Corr. 1, November 27, 1962, in U.S. Arms Control and Disarmament Agency, *Documents on Disarmament, 1962*, Publication No. 19, Vol. 2 of 2, Washington, DC: U.S. Government Printing Office, November 1963, pp. 834-852.

44. On these points, see Henry Sokolski, "Market-based Nonproliferation," testimony presented before a hearing of the

House Committee on Foreign Affairs, "Every State a Superpower?: Stopping the Spread of Nuclear Weapons in the 21st Century," May 10, 2007.

45. On these points, see Charles Ferguson, *Nuclear Energy: Balancing Benefits and Risks,* New York: The Council on Foreign Relations, April 2007.

46. See Frank von Hippel, "Managing Spent Fuel in the United States: The Illogic of Reprocessing," January 2007, an NPEC-commissioned paper published as Research Report No. 3 by the International Panel on Fissile Materials, available from *www.npec-web.org/Frameset.asp?PageType=Single&PDFFile=20070100-VonHippel-ManagingSpentFuel&PDFFolder=Essays;* Richard K. Lester, "New Nukes," *Issues in Science and Technology,* Summer 2006, pp. 39-46, available from *www.npec-web.org/Frameset.asp?PageType=Single&PDFFile=20060600-Lester-NewNukes&PDFFolder=Essays;* and Edwin Lyman, "The Global Nuclear Energy Partnership: Will It Advance Nonproliferation or Undermine It?" presented at the annual meeting of the Institute of Nuclear Materials Management, July 19, 2006, available from *www.npec-web.org/Essays/20060700-Lyman-GNEP.pdf.*

47. See U.S. Department of Energy, Office of Fuel Cycle Management, *Global Nuclear Energy Partnership Strategic Plan,* Washington, DC: U.S. Department of Energy, GNEP-167312, Rev. 0, January 2007), p. 5, where the DoE notes that "there is no technology 'silver bullet' that can be built into an enrichment plant or reprocessing plant that can prevent a country from diverting these commercial fuel cycle facilities to nonpeaceful use. From the standpoint of resistance to rogue-state proliferation, there are limits to the nonproliferation benefits offered by any of the advanced chemical separations technologies, which generally can be modified to produce plutonium. . . ."

APPENDIX I

NPEC IAEA SAFEGUARDS WORKSHOP PARTICIPANTS, 2005-2007

Graham Andrew,
Special Assistant to the Director General for Science & Technology, Office of the Director General - IAEA

Andrew Barlow,
Foreign and Commonwealth Office, United Kingdom

Patrick Beau,
National Defense General Secretariat of France

Wyn Bowen,
King's College London

Martin Briens,
Sours-Directeur Du Désarmement et De La Non-prolifération Nucléaires, French Ministère des Affaires étrangères

Lisa Bronson,
Deputy Undersecretary of Defense for Technology, Security Policy, and Counterproliferation, National Defense University

Kory Budlong-Sylvester,
NNSA - Department of Energy - USA

John Bunney,
Former Adviser to the Department of Safeguards, IAEA

Ed Burrier,
Professional Staff, Subcommittee on International Terrorism and Nonproliferation

Doug Campbell,
Office of Congressman Berman, USA

Thomas B. Cochran,
Chief Nuclear Scientist, Natural Resources Defense Council

Leland Cogliani,
Government Accountability Office

Garry Dillon,
International Atomic Energy Agency (retired)

Bruno Dupre,
Ministry of Defence, France

Jack Edlow,
President, Edlow International

Steve Elliott,
Department of State, Chief of Staff, Office of the Under Secretary of State for International Security

Phillipe Errera,
Députe Director du Centre d'Analyse et de Prevision Ministère des Affaires étrangères

Steve Fetter,
Dean, School of Public Policy, University of Maryland

Mark Fitzpatrick,
International Institute for Strategic Studies

Jeff Fortenberry,
Congressman, United States House of Representatives

Paul Fouilland,
État-major des armées, France

Thomas Göbel,
Ministry of Foreign Affairs, Germany

Jose Goldemberg,
former Brazilian Secretary of State for Science and Technology

Pierre Goldschmidt,
former Deputy Director General, Head of the Department of Safeguards, IAEA (Retired)

Bruno Gruselle,
Fondation pour le Recherche Strategique, Paris

Thomas Guibert,
Ministère des Affaires étrangères

Erwin Häckel,
German Council on Foreign Relations

Glenn Hawkins,
Department of Trade and Industry, U.K.

Brian Jenkins,
RAND Corporation

Joanna Kidd,
King's College

Bernadette Kilroy,
Director, Strategic Planning Office, Department of State, USA

Michael Knights,
The Washington Institute for Near East Policy

Michael Levi,
King's College London

Ed Levine,
Senior Professional Staff Member, Senate Foreign Relations Committee

Edwin Lyman,
Union of Concerned Scientists

Cécile Maisonneuve,
Assemblée Nationale, France

Jim McNally,
U.S. Department of State

Quentin Michel,
EU Commission, Expert National Détaché on dual use export controls

Thomas Moore,
Majority Staff, Senate Foreign Relations Committee

Raphaële Pailloux,
Delegation General for Armaments (DGA) of France

Florian Riendel,
First Secretary (Political), Embassy of Germany

Guy Roberts,
NATO, Deputy Assistant Secretary General for WMD

Henry S. Rowen,
The Hoover Institution, Stanford, former member of the U.S. Presidential Commission on he Intelligence Capabilities of the United States Regarding Weapons of Mass Destruction

Wolfgang Rudischhauser,
German Foreign Ministry

Guillaume Schlumberger,
Fondation pour le Recherche Strategique, Paris

Thomas Shea,
Pacific Northwest National Laboratory

Henry Sokolski,
Nonproliferation Policy Education Center

Pam Tremon,
U.S. Embassy, London

Robert Zarate,
Nonproliferation Policy Education Center

APPENDIX II

THE PROLIFERATION DANGERS OF LWRS

Adding to the IAEA's nuclear inspection challenges is the continued spread of large research and power reactors to countries like Egypt, Algeria, and Iran that require lightly enriched uranium as fuel and produce a significant amount of plutonium-laden spent fuel—materials, which, as has already been noted, could be seized to accelerate the production of weapons-usable uranium or plutonium. Most of these reactors are off-line fueled and so are considered to be "proliferation resistant" because their fuel cannot be removed or inserted without shutting the entire reactor down and because they are fueled with slightly enriched uranium that only a handful of advanced nuclear nations can produce. This makes inspections against possible diversions or misuse of the fuel easier than with graphite or heavy-water moderated reactors like those found in Israel, India, and North Korea where the reactor is fueled "on-line," i.e., while the reactor is still operating with natural uranium, a fuel that, unlike lightly enriched uranium, is much easier to produce indigenously.[48]

But with the development in North Korea and Iran of covert enrichment and reprocessing facilities, the proliferation resistance of even these "peaceful" reactors now is far less than advertised. In fact, one could seize all, or a portion, of the many tons of fresh lightly enriched uranium fuel that normally sits outside of most power reactors for safety reasons.[49] divert it to a covert or declared enrichment plant, open the fuel rods, crush the uranium oxide fuel pellets, heat them, and run fluoride gas over the material. The result would be the quick production of massive amounts of uranium hexafluoride without ever having to mine and mill uranium ore, or use a complicated hexafluoride production plant. More important, the enrichment of the uranium produced would reduce five-fold the amount of effort otherwise required to enrich natural uranium for use in nuclear explosives. This would significantly reduce the amount of time required for a country to produce its first uranium bomb.[50]

Yet another way that would-be bomb makers could exploit the operation of large reactors would be to divert the reactor's spent fuel either from the reactor itself, or from its spent fuel pond. Spent fuel is laden with plutonium—itself a nuclear fuel, which—once chemically stripped from the other spent fuel by-products, can make nuclear weapons of any yield. In fact, during the normal operation of large light water reactors of the sort Iran is building at Bushier, the reactor will produce 330 kilograms of near-weapons grade plutonium—enough to make over 50 crude nuclear bombs.[51]

As for chemically separating the plutonium from spent fuel, this could be accomplished in a facility as small as 65 feet by 65 feet (small enough to be built and hidden within an existing large warehouse). This plutonium separation plant also need not be elaborate. Yet another "quick-and-dirty" design plant, detailed by the nuclear industry's leading experts in the late 1970s (measuring 130 feet by 60 feet by 30 feet, see Figure 5 below), employs technology little more advanced than that required for the production of dairy products and the pouring of concrete.

Simple, Quick Reprocessing Plant Designed to Make As Many as 20 Bombs a Month (Ferguson-Culler)

10-day startup, 1 bomb's-worth-a-day production rate

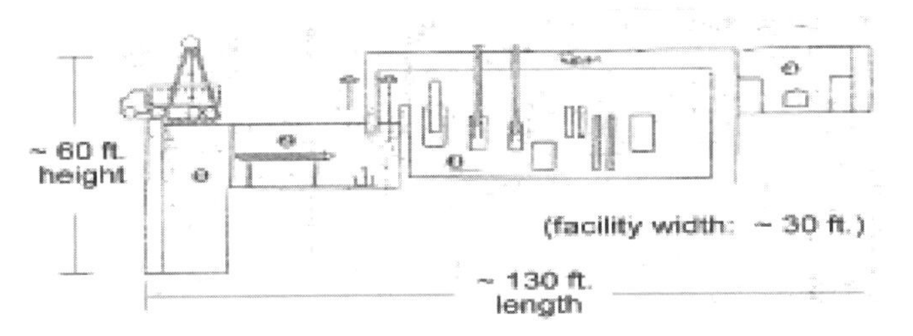

Source adapted from D.E. Ferguson "Simple Quick (Re)processing Plant" Memorandum to F.L. Gulier, Oak Ridge National Laboratory, August 30, 1977; and J.A. Hassberger, "Light-Water Reactor Fueling Handling and Spent Fuel Characteristics," Fission Energy and System Safety Program, Lawrence Livermore National Laboratory, circa February 25, 1999.

Figure 5.

These relatively compact plutonium chemical separation plants could be built within other larger buildings undetected, would not send off any signal until operated, and could separate a bomb's worth of plutonium each day after the first 10 days of operation. Assuming the country in question had already perfected a working implosion device,[52] the separated plutonium could be inserted to make a bomb directly—i.e., much more quickly than any outside party could act to block the diversion.

ENDNOTES - APPENDIX II, CHAPTER 1

1. An additional argument often offered to explain why light water reactors are proliferation resistant is that the plutonium they produce is "reactor" grade rather than "weapons" grade. This argument is specious. Reactor-grade plutonium will normally

contain about 25 percent "even isotope" plutonium (Pu 240 and Pu 242). This even isotope plutonium reduces the predictability of the precise weapons yield of any explosive device that uses it but reactor-grade plutonium can be relied upon to produce bombs with a minimum yield of at least one kiloton. Reactor-grade plutonium is also more hazardous to handle than weapons-grade plutonium, which normally contains no more than 6 percent even isotope plutonium. Still, for most national weapons efforts, the disadvantages of reactor-grade plutonium can be surmounted with proper weapons design adjustments to make a weapon of any yield. On these points, see J. Carson Mark, "Explosive Properties of Reactor-Grade Plutonium," *Science and Global Security*, 4 (1993), p. 111 and U.S. Department of Energy, *Nonproliferation and Arms Control Assessment of Weapons-Usable Fissile Material Storage and Excess Plutonium Disposition Alternatives* Washington, DC: U.S. Department of Energy, DOE/NN-0007, 1997, pp. 37-39.

2. For a large power reactor of the size of Iran's Bushier reactor, it is customary to keep one reload, a third of a core consisting of 20 tons of lightly enriched uranium fuel, at the reactor site.

3. For additional details on how fresh light water reactor fuel could be used to accelerate a uranium weapons program, see Victor Gilinsky, Harmon Hubbard, and Marvin Miller, *A Fresh Examination of the Proliferation Dangers of Light Water Reactors,* Washington, DC: The Nonproliferation Policy Education Center, October 22, 2004, pp. 35-41, reprinted in Henry Sokolski, ed., *Taming the Next Set of Strategic Weapons Threats,* Carlisle, PA: Strategic Studies Institute, U.S. Army War College, 2006, available from *www.npec-web.org/Frameset.asp?PageType=Single&PDFFile=20041022-GilinskyEtAl-LWR&PDFFolder=Essays.*

4. This near-weapons-grade material is referred to as "fuel" grade plutonium and contains no more than 14 percent even-isotope plutonium. For a detailed discussion of the weapons utility of reactor and fuel-grade plutonium as compared to weapons-grade, see, Gilinsky, *A Reassessment,* pp. 21-33; and Harmon W. Hubbard, "Plutonium from Light Water Reactors as Nuclear Weapons Material," April 2004, Washington, DC: The Nonproliferation Policy Education Center, available from *www.npec-web.org/projects/hubbard.pdf.*

5. Although developing a working implosion device that can be used with either uranium or plutonium nuclear fuel is much more challenging than perfecting a working gun device, which can only be used to make a uranium bomb, it should no longer be assumed to be a major technical hurdle for most nations. Saddam Hussein's scientists perfected a working implosion device over 15 years ago. Working, tested designs have also been shared with at least Pakistan, Israel, and Libya by the French, United States, China, and Pakistan. For more on these points, see Barton Gellman, "Iraqi Work Toward A-Bomb Reported U.S. Was Told of 'Implosion Devices'," *The Washington Post,* September 30, 1998, p. A01; Carey Stublette, "Pakistan's Nuclear Weapons Program Development," January 2002, available from *nuclearweaponarchive.org/Pakistan/PakDevelop.html*; BBC News, UK Edition, "China 'Link' to Libya Nuke Design," February 16, 2004, available from *news.bbc.co.uk/1/hi/world/middle_east/3491329.stm*; and Avner Cohen, *Israel and the Bomb*, New York, NY: Columbia University Press, 1998, pp. 82-83; and Steve Weissman and Herbert Krosney, *The Islamic Bomb: The Nuclear Threat to Israel and the Middle East,* New York: Times Books, 1981, pp.114-117.

PART II:

NEW IAEA INSPECTIONS POSSIBILITIES

CHAPTER 2

CAN WE TRACK SOURCE MATERIALS BETTER – DO WE NEED TO?

Jack Edlow

Uranium is a naturally occurring element found in low levels within all rock, soil, and water. This is the highest-numbered element to be found naturally in significant quantities on earth. It is considered to be more plentiful than antimony, beryllium, cadmium, gold, mercury, silver, or tungsten, and is about as abundant as arsenic or molybdenum. It is found in many minerals including uraninite (also called pitchblende, the most common uranium ore), autunite, uranophane, torbernite, and coffinite. Significant concentrations of uranium occur in some substances such as phosphate rock deposits and minerals such as lignite and monazite sands in uranium-rich ores (it is recovered commercially from these sources).

All of this is to say that uranium is found in most countries at least in some concentrations, and in many countries in fairly rich deposits. Uranium has been mined in many countries around the world, including Australia, Brazil, Argentina, Portugal, France, East Germany, Bulgaria, Czechoslovakia, Niger, Gabon, Namibia, South Africa, Zaire (Democratic Republic of the Congo [DRC]), Russia, United States, Canada, Kazakhstan, Uzbekistan, China, Mongolia, and Sweden. New mines are under development in Malawi, Zambia, and Uganda, to name a few.

It should be obvious that uranium, as a source material, can be used within even a small commercial research reactor to create quantities of plutonium that can in turn be used to create weapons. This could be done in such a way as to circumvent international safeguards. The case of the Osirak Reactor bombed by the Israelis on June 7, 1981 under Operation OPERA was deemed by Israel *and* Iran to be such a case. Ironically, the Iranians had bombed the reactor on September 30, 1980, but had not destroyed it. Israel was more successful. Why was it necessary to bomb the reactor? Iraq had obtained large quantities of natural uranium either through open commercial means or through stealth. This material would have been transmuted into plutonium 239 in the reactor. Both Iran and Israel felt the need to deal with the threat before it became a certainty.

Israel well understood this method because it had itself apparently followed a similar path. Under Operation PLUMBAT in 1968, the German freighter *Scheersberg A* disappeared on its way from Antwerp to Genoa along with its cargo of some 200 tons of uranium oxide (yellowcake). When the freighter reappeared in Iskenderun, a Turkish port, the cargo was missing; it had been transferred at sea to an Israeli ship. It is believed that this uranium was transferred to the Dimona facility in Israel for use in the research reactor.

More recently, A United Nations (UN) report dated July 18, 2006, said there was "no doubt" that a huge shipment of smuggled uranium 238 uncovered by customs officials in Tanzania in October 2005 was transported from the Lubumbashi mines in the Congo. A senior Tanzanian customs official said the illicit uranium shipment was found hidden in a consignment of coltan, a rare mineral used to make chips in mobile telephones. The shipment was destined for smelting in the former Soviet republic of Kazakhstan and delivered via Bandar Abbas, Iran's biggest port. It is unlikely that this cargo would have made it to Kazakhstan. It would have been diverted for use in Iran for purposes we can only suspect.

Prima facie, these cases would call for more controls over source materials, including uranium and the other principal source material, thorium. Thorium, which can also be used to produce materials suitable for weapons applications, is found in small amounts in most rocks and soils, where it is about three times more abundant than uranium and is about as common as lead. The current thorium mineral reserve estimates are shown in Figure 1.

Country	Current Thorium Mineral Reserves (in Tons)
Australia	300,000
Brazil	16,000
Canada	100,000
India	360,000
Norway	170,000
South Africa	35,000
United States	160,000
Others	95,000

Figure 1. Current Thorium Mineral Reserves Estimates.

But there are already requirements for reporting the sale and transfer of source materials from one country to another. International Atomic Energy Agency (IAEA) member states do report these shipments, and they are generally effective, as long as the parties want them to be. The question is how to enforce these requirements. It seems obvious that in the not too distant past there has been circumvention, caused by a national program disguised to evade international detection. In the case of the Tanzanian intercept, the shipment was apparently detected by equipment installed at the port under the U.S. Megaports program to detect the potential smuggling of radioisotopes along the Indian Ocean Coast. It may have been merely a coincidence that it detected the uranium ore concentrates, BUT IT DID.

Since uranium and thorium are so abundant; since it is not illegal to sell these materials; since it is easy to ship the materials and possibly to divert them; and since the materials can be used in programs to create weapons of mass destruction (WMD), it seems that additional administrative controls, while possibly helpful, cannot be relied upon to track and control these materials. Diversion will occur, when diversion is desired.

It is because of this, that *tracking* of material needs to rely on detection. In the past, railroads kept track of their rolling stock through administrative controls, and cars were lost on sidings, sometimes for months. Subsequently, an identification system using bar coding was developed so that when cars passed detectors, their last location was known. The problem was that when the cars were not moving, only their last known location was known. More recently, global positioning systems (GPS) have been incorporated into the tracking of cars and also now truck fleets. This provides for location detection even when vehicles are not moving.

In order to detect diversion, producers of uranium could incorporate some advanced technology into the shipping components. This would detect the PLUMBAT-type circumstance as long as the shipper was not a party to the diversion.

No type of "in package" device will detect the nationally-sponsored diversion like that which occurred in Tanzania last year. Presumably, no one would wish to be detected in that case. Iran planned an elaborate mechanism to evade detection, but did not count on the MegaPorts detectors. One would expect future diversions to take this into account.

The next line of defense is to render the possession of the materials harmless. Without unsafeguarded reactors or enrichment plants, the possession of source materials is meaningless. North Korea could not use the spent fuel from its reactor as long as the reactor was under IAEA safeguards. Instead, they built an undeclared enrichment facility and later quit the safeguards regime to pursue their objectives by using the reactor fuel after all. They are clearly able to obtain source materials despite the current controls, both administrative and physical.

More physical detection equipment at seaports *and* airports would be essential to detect the movement of radioactive cargos and to alert officials to potentially unknown shipments. This, unfortunately, would also trigger many alerts based on known and existing shipments. It could slow or even impede the transport of legitimate cargos as carriers and ports prohibit the shipment of cargos so as not to impact their general operations. This leads to more "delay and denial" problems.

Australia and Canada have put substantial administrative contols on their source materials. Strict conditions apply through the bilateral agreements that these countries enter into with other nations. It is certain that their materials will not be diverted to be used inappropriately. If only all countries were to take the same approach. But such is not the case at this time. The Nuclear Suppliers Group has outlined the mechanisms for the control, transfer, and retransfer of source materials, including export licenses and physical security. This is not enough, however. It has not prevented the diversion of centrifuge technology, although it may have slowed it somewhat.

Uranium and thorium, being so widely distibuted, are much easier to mine, process, and ship. The good news is that these radioactive materials can be detected, if sufficient equipment is positioned worldwide. Beyond yellowcake, of course, are the source material products of processed uranium hexaflouride. Unlike yellowcake, only a limited number of countries currently produce uranrium hexiflouride products. As a result, monitoring the production and transfer of these materials would be much more practicable than it would be for yellowcake. In fact, it was intelligence on the transfer of Chinese uranium hexaflouride to Iran in the early 1990s that helped tip off the United States and the IAEA on Iran's undeclared uranium enrichment program. This example suggests the leveraged utilty of focusing on such transfers.

In conclusion, tracking of source material through either administrative or physical controls is essential. The methods used to date have not prevented and cannot prevent diversion of these materials. Advanced technology could be useful in further detection of attempts to divert but would not be foollproof. IAEA facility safeguards are only useful when applied to all facilities within a willing country. Tracking of source materials cannot in itself prevent development of weapons, but it can be one small tool in the process to detect and slow the diversion of materials.

CHAPTER 3

NOVEL TECHNOLOGIES FOR THE DETECTION OF UNDECLARED NUCLEAR ACTIVITIES*

Nikolai Khlebnikov, Davide Parise, and Julian Whichello

INTRODUCTION

The International Atomic Energy Agency (IAEA) works to maximize the contribution of nuclear technology to human endeavors, while verifying its peaceful use. The IAEA's mission is addressed by science and technology, mobilizing peaceful applications of nuclear science and technology to developing countries; by safety and security, protecting people and the environment from harmful radiation exposure; and by safeguards and verification, preventing the further spread of nuclear weapons. In the area of safeguards and verification, the IAEA carries out inspection activities that include confirming a state's declared nuclear material (including plutonium and enriched uranium) and maintaining vigilance for evidence of undeclared nuclear material and activities. In exceptional circumstances, the IAEA may also be granted special responsibilities under United Nations Security Council (UNSC) resolutions, allowing it to search for and uncover covert nuclear weapons programs (e.g., following the 1991 Gulf War), or to conduct ongoing monitoring of disarmament (e.g., monitoring the freeze on reprocessing plutonium under the 1994 framework agreement with the Democratic People's Republic of Korea (DPRK).

*This is International Atomic Energy Agency paper IAEA-CN-148/32.

In 2004, the IAEA General Conference called upon the Secretariat to examine innovative technological solutions to strengthen the effectiveness and improve the efficiency of IAEA safeguards. Member States also agreed to provide appropriate assistance to facilitate the exchange of equipment, material, and scientific and technological information for the implementation of additional protocols. The project Novel Techniques and Instruments for Detection of Undeclared Nuclear Facilities, Material, and Activities (known as the Novel Technologies Project) was established in 2005 to identify specific needs and initiate the necessary research and development (R&D) of techniques and instruments that will be used for the implementation of additional protocols, including the conduct of complementary access.

The IAEA Strategic Objectives for 2006-11[1] include the enhancement of the IAEA's detection capabilities through the development of new or improved safeguards approaches and techniques, and the acquisition of more effective verification equipment. The following goals are applicable to the Novel Technologies Project:

- Improve current detection capability;
- Pursue R&D for the development of novel technologies for detection of undeclared activities;
- Utilize, *inter alia,* Member States Support Programme (MSSP) mechanisms as well as internal resources and expertise; and,
- Optimise safeguards equipment and technology.

DEVELOPMENT AND IMPLEMENTATION OF SAFEGUARDS METHODS AND INSTRUMENTS

Implementation of effective and efficient safeguards has increasingly relied on the development and deployment of methods and instruments meeting specific functional and technical requirements. Accordingly, equipment development has complemented the safeguards implementation approaches. For example, early safeguards equipment was developed in support of on-site verification of materials and activities at declared locations.

After the 1991 Gulf War and the discovery of a clandestine nuclear weapons program in Iraq, safeguards approaches were enhanced to include additional methods and techniques, providing the IAEA with further tools by which it could better detect undeclared activities. These included environmental sampling, information analysis, export monitoring, satellite imagery, and new technologies such as ground penetrating radar. New technologies were also developed in support of additional protocols activities, including those for complementary access.

By their very nature, clandestine weapons programs take place at undeclared locations or at declared locations that may be used as a "cover" for an undeclared process being carried out. The location of such activities requires appropriate equipment that can detect unique characteristics related to the particular activity. The Novel Technologies Project aims to broaden the range of techniques and instruments available to the IAEA, including emerging techniques and instruments that enable the IAEA to detect undeclared activities in undeclared locations (e.g., small industrial areas, universities, and workshops).

THE NOVEL TECHNOLOGIES PROJECT

In 2005, the IAEA Department of Safeguards solicited suggestions and proposals through its MSSP system. Broad requirements based on safeguards needs were prepared and sent to all MSSPs and other international bodies. Over 60 proposals, covering a wide range of techniques, were received and reviewed by the Safeguards Department. Techniques regarded as "new"[2] were forwarded to the relevant organizational unit in the IAEA for further consideration. Those regarded as "novel"[3] methods or instruments addressing a particular safeguards problem were selected for further development and evaluation within the Novel Technologies Project. Interestingly, many were based on emerging laser and other forensic techniques.

Project Tasks.

The following proposals, meeting specific safeguards needs for both on-site and away-from-site detection of undeclared activities, have been selected by the IAEA for further development and evaluation.

Optically stimulated luminescence (OSL).

Need: To determine if an undeclared location has been used previously for storing radiological material.

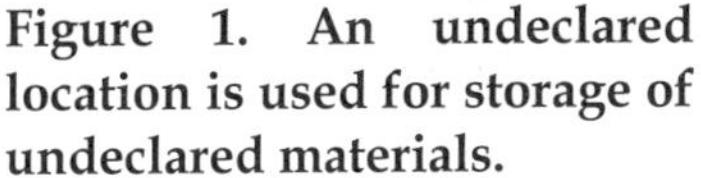

Figure 1. An undeclared location is used for storage of undeclared materials.

Figure 2. The materials are removed and the location is subsequently "disquised."

Proposed Solution: Use OSL to measure the radiation-induced signature retained in many common building materials.

Figure 3. An IAEA inspector collects samples of the surrounding building materials.

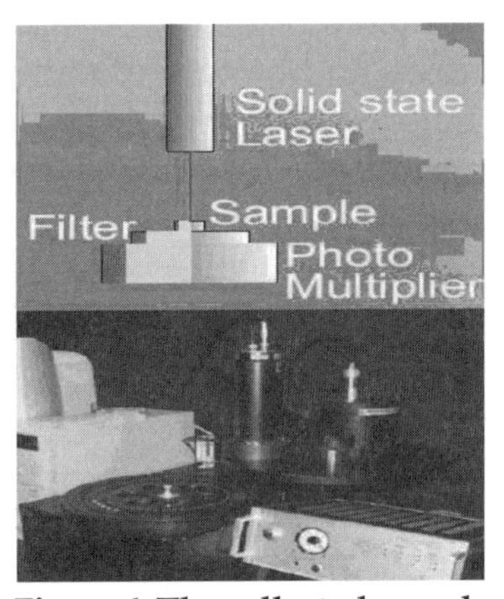

Figure 4. The collected samples are analyzed for residual nuclear activation, indicating the previous presence of stored nuclear materials.

Laser-induced breakdown spectroscopy (LIBS).

Need: To determine the nature and history of compounds and elements found on site.

Figure. 5. Unidentified materials found during an on-site complementary access inspection.

Proposed Solution: Use on-site LIBS to determine the nature and history of compounds and elements.

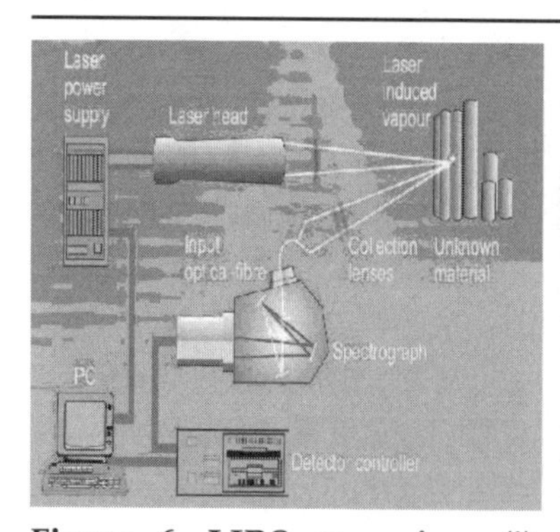

Figure 6. LIBS comprises (i) a laser system to ablate the material surface to create a micro-plasma, and (ii) a spectrometer to generate a spectroscopic profile of the micro-plasma's constituent components.

Figure 7. A trained IAEA inspector operates the LIBS unit on-site. The spectroscopic profile is compared to those in its library to determine material's make-up and history.

Light detection and ranging (LIDAR).

Need: To detect the presence and nature of nuclear fuel cycle process activities at suspected locations.

Proposed Solution: Use a mobile LIDAR laboratory in the vicinity of a suspected site to detect the presence of characteristic gaseous compounds, emanating from nuclear fuel cycle (NFC) processes into the atmosphere.

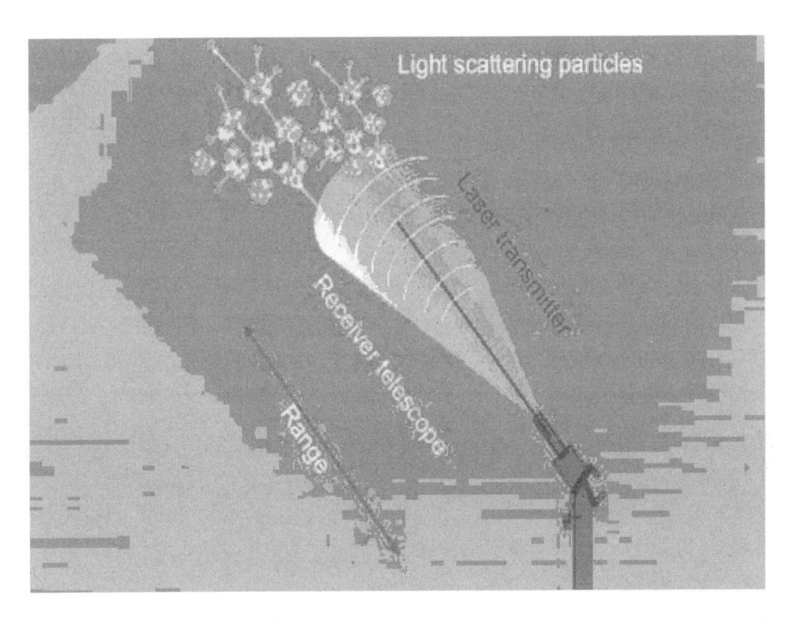

Figure 8. LIDAR methods are used routinely by environmental monitoring agencies to determine the presence of pollutants in the atmosphere.

Figure 9. A mobile LIDAR travels to the vicinity of a suspected location engaged in undeclared NFC processes. A laser, tunable to precise wavelengths (λ) selectively stimulates specific airborne molecules emanating as gaseous compound from the process. A light-sensitive telescope scans the atmosphere, detecting the presence of the stimulated molecules.

Sampling and analysis of atmospheric gases.

Need: To detect the presence and nature of nuclear fuel cycle process activities at suspected locations.

Proposed Solution: Use on-site laboratory to determine the atmospheric composition of gaseous mixtures.

Figure 10. A mobile on-site laboratory samples and concentrates atmospheric-borne pollutants. Local meteorological conditions and the GPS location are also recorded.

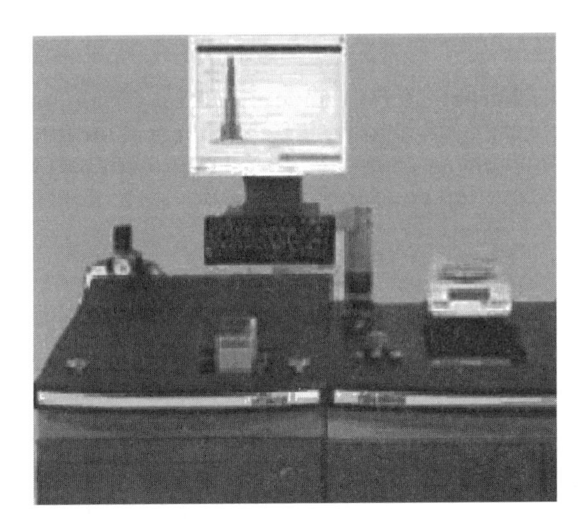

Figure 11. Samples are brought to a field laboratory for analysis.

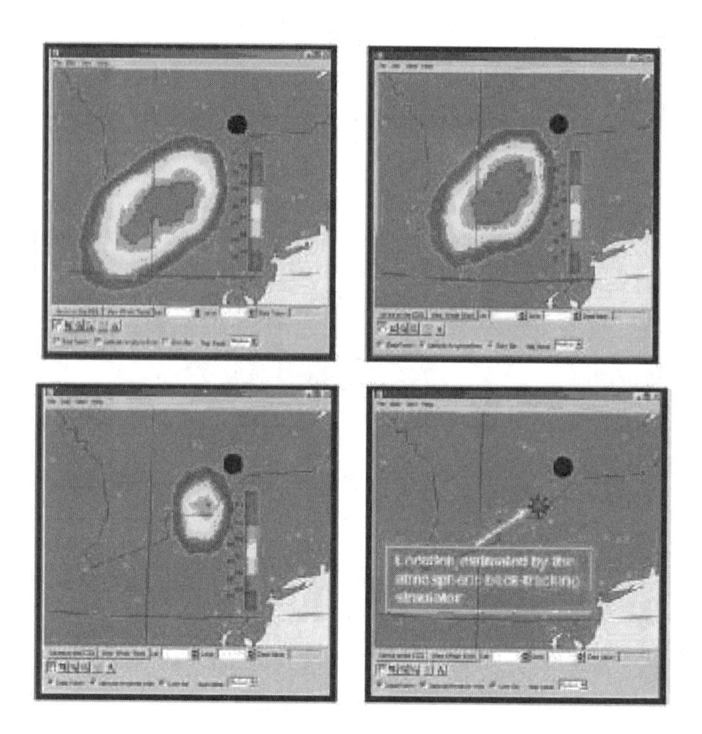

Figure 12. Airborne material is identified through sample analysis, and the data are combined with meteorological information in a suitable atmospheric computer model to provide an estimate of the source direction and probable location.

Project Activities.

In parallel to pursuing the tasks outlined in Project Tasks, the Project has also convened specialist technical meetings on techniques for the verification of enrichment activities,[4] noble gas sampling and analysis,[5] and laser spectrometry techniques.[6] Further specialist meetings covering novel technologies are being planned. Additionally, the Project has been active with the support of Member States in establishing contacts with international R&D organizations and with experts engaged in a wide range of sensor and detection technologies. MSSPs have also been supportive, agreeing to assist the Project by facilitating technical exchanges with both private and government-operated R&D laboratories and by providing access to experts for short-duration tasks, facilitating attendance at technical meetings, advising on novel methods and instruments, conducting field tests and providing supplementary funding.

The Project is also developing a secure technical database to handle relatively large volumes of technical information. The database will also provide nonsensitive information on the Project's tasks and activities on a dedicated website to further raise the profile of this work to the international R&D community.

Project Planning.

The Novel Technologies Project was established to develop and evaluate effective techniques that meet IAEA needs and that can be incorporated within safeguards approaches for detecting evidence of undeclared nuclear fuel-cycle activities, particularly at undeclared locations. To that end, the Project will continue to conduct surveys to identify safeguards needs that cannot be met with available techniques, broaden technical collaboration with other nonproliferation organizations and the international R&D community and, where required, initiate further tasks that will lead to safeguards-useable methods and instruments. The basis of these initiatives will be a review and analysis of the nuclear fuel cycle processes, the identification of the most safeguards-useful activity indicators[7] and emanating signatures[8] that can "travel" from the source location and be detected with a high level of confidence and accuracy. Indicators and signatures will be information, matter, and/or energy associated with a particular NFC process. Once identified, methods useful for the detection of promising indicators, and signatures will be assessed by experts to determine if suitable methodology or instruments are available. Where none exist in a safeguards-useable form, then the Project will define appropriate technical and procedural requirements, initiating the necessary R&D and testing regimes.

CONCLUSIONS

The establishment of the Novel Technologies Project has provided a mechanism for the IAEA to address the technologies required for emerging and future inspectorate needs. Moreover, it has facilitated the IAEA's access to a greatly expanded range of methods and instruments, thereby allowing safeguards planners the opportunity to develop novel verification and detection approaches.

REFERENCES

International Atomic Energy Agency, *Technical Meeting on Techniques for IAEA Verification of Enrichment Activities, STR-349,* Vienna: International Atomic Energy Agency, 2005.

International Atomic Energy Agency, *Strategic Objectives 2006–2011,* Vienna: International Atomic Energy Agency, 2006.

International Atomic Energy Agency, *Technical Meeting on Application of Laser Spectrometry Techniques in IAEA Safeguards,* Vienna: International Atomic Energy Agency, 2006.

International Atomic Energy Agency, *Technical Meeting on Noble Gas Monitoring Sampling and Analysis for Safeguards,* Vienna: International Atomic Energy Agency, 2006.

ENDNOTES - CHAPTER 3

1. International Atomic Energy Agency, *Strategic Objectives 2006-2011*, Vienna: International Atomic Energy Agency, 2006.

2. New technologies are defined as those for which the methodology is already understood and implemented by the IAEA for safeguards applications. Examples include the next generation surveillance and sealing system.

3. Novel technologies are defined as those for which the methodology has not been applied previously by the IAEA for safeguards applications.

4. International Atomic Energy Agency, *Technical Meeting on Techniques for IAEA Verification of Enrichment Activities, STR-349*, Vienna: International Atomic Energy Agency, 2005.

5. International Atomic Energy Agency, *Technical Meeting on Noble Gas Monitoring Sampling and Analysis for Safeguards*, Vienna: International Atomic Energy Agency, 2006.

6. International Atomic Energy Agency, *Technical Meeting on Application of Laser Spectrometry Techniques in IAEA Safeguards*, Vienna: International Atomic Energy Agency, 2006.

7. Indicators are defined as entities that go into making the process operative. Examples are resources, required materials, facility design and related R&D.

8. Signatures are defined as entities produced by the nuclear fuel cycle process when it is in operation Examples are produced material, process by products and energy emanations.

CHAPTER 4

WIDE AREA ENVIRONMENTAL SAMPLING IN IRAN

Garry Dillon

In 2005 NPEC commissioned me to write this paper on WAES in Iran which I did - after many false starts. The paper was not meant to be any more than a starting point. It contains many gross assumptions and simplifications. There are a number of costs not addressed such as internal transport and security measures for the sampling stations. My rationale was that the overall cost would be so influenced by the actual detection range and practical servicing frequency that the "odd million here or there" would be of little consequence.

Any meaningful negotiated agreement with Iran must, at the very minimum, require Iran to:

1. Unconditionally agree not to acquire or develop nuclear weapons or weapons -useable nuclear material or any subsystems or components or any research, development, support or manufacturing facilities related thereto.

2. Submit to the Secretary General and to the Director General of the International Atomic Energy Agency (IAEA) a declaration containing complete details of its past activities to produce or acquire special nuclear materials (plutonium and enriched uranium) and nuclear weapons technology, including complete details of external assistance, both offered and provided, as well as related procurement activities.

3. Immediately enter into force the "Additional Protocol" and actively cooperate with the IAEA in its robust implementation for the purpose of verifying the accuracy and completeness of the information provided with respect to Iran's past nuclear activities and in building ongoing confidence that Iran's present and future nuclear activities are confined to exclusively peaceful uses.

The IAEA is very well-experienced in the verification of the accuracy and completeness of declarations and, while significant uncertainties remain, has already gained considerable knowledge of Iran's past nuclear activities.

The IAEA has also gained much experience in Wide Area Environmental Sampling (WAES), both through its field and laboratory work in laying the technical groundwork for the Additional Protocol and from the implementation of its plan for the Ongoing Monitoring and Verification (OMV)[1] of Iraq's undertaking in compliance with paragraph 12 of United Nations Security Council (UNSC) Resolution 687 (1991).

This chapter addresses the viability of WAES—as defined in Article 18.g of the Additional Protocol—as a means of gaining assurance that all nuclear activities in Iran are known to the IAEA and are subject to routine verification in accordance with Iran's safeguards agreement with the IAEA.

The context for the implementation of WAES in Iran would be twofold:

- Iran is already a potential nuclear weapons state, and should Iran so choose, it would be merely a matter of time until it fully developed a production capability for weapons-usable nuclear material and its subsequent weaponization.
- Iran's oft-declared peaceful nuclear undertakings might well be genuine, but pragmatism requires that comprehensive verification measures be implemented over a substantial period of time to build the necessary level of international confidence of Iran's compliance with those undertakings.

The Locations and Scope of Monitoring Activities Prescribed by the Additional Protocol.

The Additional Protocol defines, inter alia, the locations to which the IAEA may gain "Complementary Access"[2] and specifies the verification measures the IAEA may implement to gain assurance of the absence of nuclear material or nuclear-related activities. The locations fall into three specific categories:

1. Locations not holding, or no longer holding, nuclear material.

2. Locations holding material that has not been processed to a level of purity for it to be suitable for fuel fabrication or isotopic enrichment, or holding material that has been exempted from safeguards verification measures by virtue of its non-nuclear use or nuclear material (typically in the form of waste) judged to be unrecoverable.

3. Locations hosting nuclear-related activities but not holding nuclear material.

The verification measures for the three categories comprise a subset of the following:

- Visual observation (common to all three categories).
- Collection of environmental samples (common to all three categories).
- Radiation detection and measurement (common to all three categories).
- Nondestructive measurements and sampling.
- Item counting.
- Application of seals...and other tamper indicating devices.
- Examination of records...of [nuclear] material.
- Other objective measures...agreed by the [IAEA] Board.

These locations and verification activities are summarized in Table 1.

SUMMARY OF LOCATIONS SUBJECT TO COMPLEMENTARY ACCESS AND PERMITTED VERIFICATION MEASURES

Main para	Sub para	Generic description		Permitted verification measures
		Locations not holding or no longer holding nuclear material		
5.a (i)		Any place on a *site* (co-located with a declared nuclear facility)	6.a	visual observation
5.a.(iii)		Any decommissioned facility or location outside facility (LoF)		collection of environmental samples
				radiation detection and measurement
				seals
				other objective measures - BoG approved
		Locations holding material that has not been processed to a level of purity for it to be suitable for fuel fabrication or isotopic enrichment or holding material which has been exempted from safeguards by virtue of its non-nuclear use or nuclear material for which safeguards has been terminated (Waste)		
5.a (ii)	2.a (v)	Any uranium mine or uranium or thorium concentration plant	6.b	visual observation
	2.a (vi) (a)	Any location holding source material quantities exceeding 10Te uranium and/or 20Te thorium		collection of environmental samples
		and other locations holding more than 1Te of such material where the collective holding of all such		radiation detection and measurement
		other locations exceeds 10Te uranium or 20Te thorium respectively		examination of records
	2.a (vi) (c)	Any location of the import of uranium and thorium in the above actual or cumulative quantities		(material quantities, origin and disposition)
	2.a (vii) (a)	Any location holding exempted material		NDA and sampling
	2.a (vii) (b)	Any location holding exempted material not yet in non-nuclear use form		item counting of nuclear materials

	2.a (viii)	Any location storing or processing high-level waste on which safeguards have been terminated		other objective measures - BoG approved
		Locations hosting nuclear- related activities but not holding nuclear material		
5.b	2.a.(i)	Any location of nuclear fuel cycle-related activities not involving nuclear material	6.c	visual observation
	2.a.(iv)	Any location engaged in Annex I activities		collection of environmental samples
	2.a.(ix) b	Any location holding imported Annex II materiel		radiation detection and measurement
	2.b	As 2.a (i) but where activities are not state funded or carried out on behalf of the state		examination of safeguards relevant records
				(production and shipping records)
				other objective measures - BoG approved
5.c		**Any location specified by the IAEA to carry out location specific environmental sampling**	6.d	collection of environmental samples
				AND IF QUESTION UNRESOLVED
				visual observation
				radiation detection and measurement
				other objective measures - BoG approved

Table 1. Summary of Locations Subject to Complementary Access and Permitted Verification Measures.

It is clear that, under the Additional Protocol, the IAEA's right of (complementary) access and freedom of choice of verification activities are considerably less than was provided in the case of its OMV plan implemented in Iraq. Nonetheless, given Iran's active cooperation, there is enough flexibility in the text of the protocol for the IAEA to be able to implement a verification process that would provide substantial assurance of Iran's compliance with its undertakings.

In the special context of this chapter, two Articles of the Additional Protocol are of fundamental importance:

1. Article 5.c, which would require Iran to ". . . provide the Agency [IAEA] with access to any location specified by the Agency . . . to carry out *location-specific environmental sampling . . .,*" and

2. Article 9, which would require Iran to ". . . provide the Agency with access to locations specified by the Agency to carry out *wide-area environmental sampling . . .*"

Article 9, however, goes on to state that "The Agency shall not seek such access until the use of *wide-area environmental sampling* and the procedural arrangements therefore have been approved by the Board and following consultations between the Agency and [the state]."

Wide-area environmental sampling is defined in Article 18.g as meaning the collection of environmental samples (e.g., air, water, vegetation, soil, and smears) at a set of locations specified by the Agency for the purpose of assisting the Agency to draw conclusions about the absence of undeclared *nuclear material* or nuclear activities over a wide area. *Location-specific environmental sampling* differs only in its application being confined ". . . at, and in the immediate vicinity of a location . . ." and that the Agency's conclusions are drawn with respect to that ". . . specific location . . ."

It is clear from the foregoing that a legal basis for the implementation of WAES in Iran exists and, with the approval of the IAEA Board of Governors and the cooperation of Iran, could be implemented. Furthermore, the text of the Additional Protocol underwrites the fundamental value of environmental sampling as a contributing technology to the IAEA's ability to draw conclusions regarding the absence of undeclared *nuclear material* or nuclear activities. However, it is clear that a prerequisite to the implementation of WAES is a reasonable understanding of the costs involved and the technical resource requirements.

A Notional Plan for the Implementation of WAES in Iran.

There is no particular complication to the "front-end" of environmental sampling, it is simply a matter of determining what kind of sample is most appropriate and at which and how many locations the samples should be taken. The premise on which WAES is based is that any significant activities related to the processing of nuclear material would result in a detectable impact on the environment—either from chronic low-level releases or an acute high-level release following a processing malfunction.

In the context of a clandestine enrichment facility, WAES would be focused mainly on the detection of uranium but also on related processing elements such as fluorine. Due to its ubiquity, the mere detection of uranium is of little significance unless it is detected in concentrations markedly different from those occurring naturally in the area sampled or the relative abundance of the ^{235}U isotope is greater than .71 percent.

WAES is further complicated by the fact that in a state such as Iran with a history of uranium processing activities, analysis of deposition samples (such as surface smears or vegetation) would not be able to readily distinguish whether the material deposited was a result of current nuclear activities or originated from past activities. Although vulnerable to resuspension complications, it is now widely accepted that the sampling of air for the collection of particulate and gaseous matter is the most reliable and unambiguous means of detection of current nuclear activities. A variety of designs of air samplers exist, ranging from those little more sophisticated than a domestic vacuum cleaner to those capable of continuous analysis of the collected matter combined with the capability to transmit the results of that analysis to a headquarters control room.

However, the IAEA's experience in Iraq shows that simplicity of design and robustness of construction are likely to provide the most reliable performance. Ideally, the air sampling equipment would be housed within a small trailer or road vehicle and would have battery back-up and the capability to transmit alarm annunciations to an appropriate IAEA control and supervision location in the event of loss of power supplies or tampering. The transportability of the air sampling station enables the grid to be readily adjusted or, if appropriate, completely redesigned.

Nuclear forensics have achieved such extraordinary sensitivities that it is virtually impossible to sanitize radioactively contaminated surfaces or to avoid the detection of leakages of radioactive airborne or liquid discharges. For example, analysis of environmental samples—airborne particulate matter, water, deposited or sedimented materials—is capable of detecting the presence of uranium down to a few millionth, billionth, billionth parts of a gram. However, even with such sensitivities, it has to be recognized that the concentration of any environmental contamination reduces inversely and nonlinearly with the distance from the point of release. The actual reduction would be a function of terrain and the prevailing meteorological conditions.

A detailed topographical/meteorological study would thus be required to determine a practical detection range based on an assessed notional release and the practical limits of sample analysis. It is, of course this "detection range-R" that will determine the grid array of the air sampling stations and thus the related capital equipment costs, in-field service personnel resources costs, and analytical costs.

For purposes of this notional plan, it is assumed that air sampling stations would be set up on a regular square grid with the diagonal separation of the air sampling stations equal to the detection range. It follows that the grid would be of side $R/\sqrt{2}$ with a corresponding grid-element area of $R^2/2$. As a first approximation the number of air sampling stations could be calculated by dividing Iran's superficial area by the area of a single grid element. However, it would also be necessary to recognize the need for additional sampling stations to cover the perimeter of the grid. Again, for purposes of illustration, the number of additional air sampling stations is derived from the state's land boundary divided by $R/\sqrt{2}$.

Iran is not a small country and covers an area of some 1.65 million square kilometers with a land boundary of 5,400km. Based on the foregoing assumptions, a sampling network designed around a detection range of 10km would require an unmanageable 33,764 air sampling stations and even stretching the detection range to 100 kilometres would still require some 400 air sampling stations. As was the case in Iraq, it is likely that the relatively high levels of atmospheric dust would require frequent sample changing to avoid blockage of the collecting media. Assuming, therefore, that samples were changed on a bi-weekly basis, a 400-station network would generate some 10,400 samples per year.

A cost assessment of a 400-station network is shown in Table 2.

Operation	Cost
Detection range (km)	100
Number of air sampling stations	400
Equipment and installation cost per unit	$10,000
Amortization period (years)	3
Equipment and installation costs per year	$1,333,000
Number of service visits/year per installation	26
Servicing capacity (units/day-2 person team)	3
Servicing resources required (person-years[1])	40
Field Office resource requirements (person-years)	4
Personnel costs including travel and accommodation	$5,000,000
Number of samples collected for analysis	10,400
Cost per analysis	$1,000
Total analytical costs	$10,400,000
Total annual costs including equipment amortization	$16,733,000
Notional overall annual cost/air sampling station	$41,833
Notional overall cost per analytical result	$1,610

Table 2. Notional Cost Assessment of a 400-Air Sampling Station Network.

The data in Table 2 are produced simply to illustrate operational costs and should not be interpreted to suggest that the exampled network is capable of providing meaningful detection sensitivities. Indeed, the mountainous nature of much of the Iranian terrain will complicate the country-specific topographical and meteorological study and is likely to indicate the need for a nonuniform grid including areas requiring a more closely spaced grid.

Regardless of these complications, the table does show that the most critical component of overall cost is sample analysis, contributing, as it does to more than 60 peercent of the costs in the illustrative model. Furthermore, the data do not include the complementary environmental samples (herbage, smears, water, etc.) that should be collected. Although these samples could be collected without additional labor or equipment costs, they could potentially more than double the analytical load, pushing the total annual cost towards $30 million. Even without this extra burden, the number of samples generated in the Table 2 example far exceeds the currently available international analytical resources at the very highest level of sensitivity.

It should also be recognized that at $30 million these notional costs represent about 25 percent of the total annual operating budget of the IAEA Department of Safeguards, including voluntary contributions from motivated member states. At first glance, such costs seem inordinately high. However, it is merely necessary to change the comparator to, for example, the annual cost of the military action in Iraq or the "replacement costs" of Manhattan following the explosion of a 50 kiloton device to make the costs appear to be an entirely worthwhile investment.

The notional case outlined in Table 2 will clearly contain inaccuracies and is presented merely to arrive at an "order of magnitude" costing of WAES in the Iran context. However, many obvious refinements are available.

One such refinement would be to carry out a detailed analysis of Iran on the basis of a 10km grid, and to weight each grid section with respect to any attributes therein that could contribute to sustaining clandestine nuclear activities. Such attributes would include, for example, access to power and water supplies, population centers, road and rail transport, and geological conditions compatible with undergrounding. On the basis of this analysis, it would be possible to determine those "high-potential" areas of the country worthy of continuous and intensive monitoring activities—probably less than 10 percent of the total area. Within the so-termed high-potential areas, air sampling stations would be positioned in conformity with location-specific detection range calculations. Areas of significantly lesser potential would be subject to less intensive sampling and analysis.

Another simple refinement is available in the analytical process in that portions of samples from contiguous locations could be blended and analyzed as a composite batch sample. Should analytical results from the composite sample so indicate, the individual sample portions could then be analyzed. It should also be recognized that WAES serves, at least in part, as a deterrent and provided the "target" is unaware, its effectiveness is undiminished regardless of whether all or only a fraction of the samples collected are actually analyzed.

Yet another refinement would be to either complement or entirely replace WAES by multiple location-specific environmental sampling wherein the locations would be chosen of the basis of their high-potential to support undeclared nuclear activities or on the basis of information provided to or independently developed by the IAEA. It is clear that motivated member states should be investing considerable resources in gathering information relevant to Iran's professed peaceful uses of nuclear energy; for example, by aerial/satellite surveillance, telecommunications monitoring, and export/import monitoring. It is equally clear that those states should be in a position to provide "cues" to the IAEA to identify *locations worthy of location-specific environmental sampling* and, as appropriate, *complementary access*.

Recommendations.

1. If not already "work in progress," the IAEA should commission a working group of internal and external experts to design a plan for the implementation of WAES in Iran based on a detailed analysis of the topographical and meteorological characteristics of its various regions. The plan should include realistic cost analyses and address the various options available between full-scope WAES and the targeted/cued implementation of multiple *location-specific environmental sampling* campaigns.

2. Motivated IAEA member states should reevaluate their relevant information gathering system and establish formal pathways for the prompt transmission of information to the IAEA.

3. Those states having high sensitivity analytical capability—principally the United States, the United Kingdom, and France—should invest significantly in the expansion and further development of those capabilities to ensure that the international community will be in a position to satisfy the demand in the event that it becomes necessary/appropriate to implement WAES in Iran or elsewhere.

4. The IAEA Board of Governors should address the question of funding for the implementation, as necessary, of WAES. Too often in the past, too many

IAEA Member States have been more focused on the financial savings that could result from the evolution of safeguards technologies and approaches. It is time to recognize that the cost of international nuclear material safeguards is trivial when compared to the financial burden of pragmatic "worst case scenarios" that might result from failure to implement robust safeguards measures at the leading edge of technological excellence.

ENDNOTES

1. As approved in United Nations Security Council Resolution 715, October 11, 1991.

2. Access to locations other than those containing declared nuclear materials.

3. Assuming a 3-month tour of duty and a 6-day working week and taking into account annual leave, official holidays, and compensatory time off, 1 person-year equates to 170 inspection days.

PART III:

SAFEGUARDS LIMITS AND PREMISES

CHAPTER 5

CAN NUCLEAR FUEL PRODUCTION IN IRAN AND ELSEWHERE BE SAFEGUARDED AGAINST DIVERSION?

Edwin S. Lyman

Introduction: Material Accountancy is Still Relevant.

The challenges to the nonproliferation regime over the last 15 years posed by the crises in Iraq, North Korea, and Iran have led to an increased preoccupation among the international community with the lack of capabilities of the International Atomic Energy Agency (IAEA) to detect undeclared facilities for production of fissile material. However, the foundation of IAEA safeguards remains the ability of the Agency to effectively verify the absence of diversion of special nuclear material from declared facilities. One must assume that the vast quantities of weapon-usable plutonium flowing through commercial reprocessing and mixed oxide (MOX) fuel fabrication plants will continue to present attractive targets to those looking to covertly acquire small stockpiles of nuclear explosives. Likewise, the huge separative work unit (SWU) capacity of large commercial gas centrifuge plants will provide a temptation for those who may wish to divert a small fraction of that capacity toward highly-enriched uranium (HEU) production. Consequently, such activities should be forbidden in the absence of highly credible assurances that all significant diversions will be detected in a timely manner. The nuclear industry will rightly not be able to increase public confidence in the security of the nuclear fuel cycle if it continues to operate facilities where dozens of bombs' worth of plutonium or HEU could conceivably go missing annually without being detected.

However, experiences with safeguarding plutonium bulk-handling facilities in Japan and Europe have made clear that, even when discrepancies in material accountancy arise, the response is anything but timely. The Agency's reluctance to escalate the significance of unresolved discrepancies to the level of violations of safeguards agreements have led to standoffs in which anomalies have remained unresolved for years or even decades. Clearly, this state of affairs is intolerable in the context of the current global threat environment.

The question of whether bulk-handling uranium facilities for conversion or enrichment can be effectively safeguarded against diversion raises somewhat different issues than those at plutonium bulk-handling facilities. Since the facilities under normal operating conditions do not involve weapon-usable process materials, the risks associated with diversion are indirect and are related to the effectiveness of enhanced safeguards measures, both to exclude the possibility of reconfiguring declared centrifuge plants to illicitly produce highly enriched uranium and to exclude the existence of clandestine enrichment plants that could utilize undeclared feed. However, even the IAEA Director of Safeguards has conceded that the additional authority provided to the Agency under the Additional Protocol is not sufficient to ensure that it will be able to discover all undeclared activities at undeclared locations.[1] Thus again, the credibility of safeguards remains dependent on the ability of international inspectors to ensure that significant quantities of nuclear materials cannot be diverted without detection from safeguarded facilities to undeclared ones, even if the materials are not direct-use.

Another reason why detection of diversion remains crucially important is the growing threat that sophisticated subnational groups, perhaps with state assistance, could obtain fissile materials to construct crude nuclear weapons for use in terrorist attacks. The world has only begun to fully appreciate the magnitude and seriousness of this danger in the aftermath of the September 11, 2001 (9/11) attacks. The potential for clandestine diversion by a state of a few significant quantities of plutonium is perhaps not the greatest proliferation concern for states that already have nuclear weapons or have large fuel cycle facilities that could be overtly commandeered for the rapid production of fissile material. But such a diversion would pose a major threat if it were carried out by or on behalf of a subnational group whose objective is to acquire only a small number of weapons for terrorist purposes. And in the latter context, the notion of timely warning as applied to states may not be relevant, since the concept as it applies to states is not directly applicable to terrorist groups that are immune to political pressures and may be able to evade capture for long periods of time if they are able to successfully escape with diverted material. Thus a security and safeguards posture that is stringent enough to deter diverters must be a fundamental goal, because the game may well be over once sufficient material is diverted.

Of course, the IAEA does not have formal authority to address subnational threats, which are the responsibility of State Systems of Accounting and Control (SSACs). However, it is apparent that improving the quality of SSACs of sufficient quality to provide the IAEA with a stringent capability to detect diversions by the operator would also provide the operator with an enhanced capability to detect diversions by insiders.[2] The problem is that aspects of domestic security that are important in countering internal threats, such as access authorization programs, would remain out of the IAEA's formal domain, even under the provisions of the revised Convention on Physical Protection of Nuclear Material (CPPNM). This dichotomy between state and nonstate actors, which appears more and more artificial today in a world where their interests are often intertwined, will hinder efforts to build comprehensive systems to effectively ensure that civil nuclear facilities cannot become covert sources of fissile material for either states or subnational groups.

There are indications that instead of moving to strengthen material accountancy practices, the IAEA is actually moving to weaken them. According to the IAEA 2004 Safeguards Statement, the Standing Advisory Group on Safeguards Implementation (SAGSI) "found that the Safeguards Criteria were basically sound, but that a key priority is the wider implementation of integrated safeguards."[3] Integrated safeguards is an effort by the IAEA to save money by reducing reliance on the results of facility-level material accounting for making state-level safeguards conclusions. However, for the reasons stated above, the IAEA should avoid an excessive focus on pursuing integrated safeguards at the expense of improving basic material accountancy measures at declared facilities.

The Challenges of Detecting Diversion at Plutonium Bulk-Handling Facilities.

In 1990, the Washington, DC-based Nuclear Control Institute issued a seminal paper by Dr. Marvin Miller of the Massachusetts Instititute of Technology, entitled "Are IAEA Safeguards on Bulk-Handling Facilities Effective?"[4] This paper illustrated, in simple yet stark terms, that the IAEA, as a result of technical and political obstacles, was unable to meet its detection goals for large-throughput plutonium bulk-handling facilities, e.g., reprocessing and plutonium fuel fabrication plants. Miller argued that this conclusion is significant because he believed that it was reasonable to regard the detection goals as performance criteria for effective safeguards, not only to maintain the credibility of the international safeguards system, but also to help ensure that national systems of accountancy and control would be stringent enough to deter subnational diversion.

One purpose of the present paper is to revisit Miller's arguments in light of any technical and political developments related to nuclear material accountancy over the last 15 years and to assess whether the conclusions of his article remain true today.

The IAEA "detection goals" have not changed since Miller's paper was written, although they are no longer universally applied. The goal remains the detection of a diversion of a "significant quantity" (SQ) of unirradiated direct use nuclear material (8 kilograms of plutonium or 25 kilograms of uranium-235 contained in HEU) within 1 month; one SQ of irradiated direct use material (about the equivalent of two pressurized-water reactor spent fuel assemblies) within 3 months; and indirect use material (75 kilograms of uranium-235 contained in low-enriched or natural uranium) within 1 year. However, these timeliness detection goals may be extended in states that have adopted the Additional Protocol and where the IAEA has concluded that undeclared nuclear materials and activities are absent, as part of the initiative known as "integrated safeguards."

Miller observed that for large bulk handling facilities, such as the 800 metric ton heavy metal (MTHM)/year Rokkasho Reprocessing Plant (RRP) now undergoing startup testing in Japan, it was not possible with the technologies and practices available at the time to detect the diversion of 8 kilograms of plutonium (1 SQ)—about 0.1 percent of the annual plutonium throughput—with a high degree of confidence. This is because the errors in material accountancy measurements at reprocessing plants were typically on the order of 1 percent—that is, a factor of 10 greater than an SQ. If after taking a physical inventory, the value of plutonium measured was less than expected (on the basis of operator records) by an amount on the order of 1 SQ, it would be difficult to state with high confidence that this shortfall, known as "material unaccounted for" or MUF, was due to an actual diversion and not merely measurement error.

In the past, the IAEA acknowledged that the 1 SQ detection goal could not be met in practice, and instead adopted a relaxed standard known as the "accountancy verification goal" (AVG), which was "based on a realistic assessment of what then-current measurement techniques applied to a given facility could actually detect."[5] The AVG was based on a quantity defined as the "expected accountancy capability," E, which is defined as the "minimum loss of nuclear material which can be expected to be detected by material accountancy," and is given by the formula $E = 3.29\sigma A$, in which σ is the relative uncertainty in measurements of the plant's inputs and outputs, and A is the facility's plutonium throughput in between periodic physical inventories.[6] This formula is derived from a requirement that the alarm threshold for diversion be set at a confidence level of 95 percent and a false alarm rate of 5 percent.

Miller estimated that for the RRP, based on an input uncertainty of ±1 percent (which was the IAEA's value at the time for the international standard for the expected measurement uncertainty at reprocessing plants), the value of E would be 246 kilograms of plutonium, or more than 30 SQs, if physical inventories were carried out on an annual basis, as was (and is) standard practice. This means that a diversion of plutonium would have to exceed this value before one could conclude with 95 percent certainty that a diversion had occurred, and that the measured shortfall was not due to measurement error.

Apparently, the IAEA no longer uses the AVG as a standard for material accountancy, and the term was not mentioned in the revised Safeguards Criteria issued in 1991.[7] The term also does not appear in the 2001 edition of the IAEA Safeguards Glossary. While some have characterized the elimination of this criterion as an attempt to strengthen material accountancy standards, it could also be regarded as a way of concealing the embarrassingly poor capabilities of conventional material accountancy methods.

Miller also identified other problems that contribute to the difficulty of detecting diversions on the order of 1 SQ, such as the accumulation of plutonium in waste streams such as cladding hulls that are not amenable to accurate assay by nondestructive means. The accumulation of plutonium in such hard-to-measure forms can lead to significant and growing contributions to cumulative facility MUFs.

An Aside on False Alarm Rates.

Even if the IAEA were able to meet its detection goals, those goals are arguably inadequate, given the evolving trends in the threats from nuclear proliferation and nuclear terrorism that have become especially apparent since the 9/11 attacks. For instance, the detection probability guidelines of 90 to 95 percent confidence level and 5 percent false alarm rate established by the SAGSI in the 1970s do not appear to be sufficiently stringent today. The Agency's reluctance to pursue higher confidence levels for detection of diversion at the expense of higher false alarm rates would seem to be a lesser concern in the context of the heightened security levels that have become standard operating practice around the world since the 9/11 attacks. Today most people are willing to tolerate a level of sensitivity for security screening at airports and critical facilities that would not have been acceptable in the past because of a common appreciation that the occasional false alarm is an appropriate price to pay to minimize the risk of another 9/11-scale terrorist attack. But the guidelines for probability of detection of diversion of plutonium have not been similarly strengthened in the aftermath of 9/11. On the contrary, ample evidence that material accountancy techniques cannot meet current quantitative detection goals does not bode well for the prospect of developing techniques capable of meeting more stringent goals without raising the acceptable false alarm rate.

Failures of Material Accountancy.

Since the release of Miller's paper in 1990, numerous examples have come to light of serious lapses in material accountancy at bulk-handling facilities around the world involving the occurrence of large MUFs that remained unresolved for years or even decades. The reasons for these lapses illustrate some of the fundamental problems encountered at bulk-handling facilities that prevent timely closure of material balances and that must be overcome if the IAEA detection goals are to be met. These problems include accumulation of residual holdup, accumulation of scrap and waste materials in hard-to-assay material forms, inaccuracies in nuclear material estimation methods, and operator complacency/incompetence.

The problem of *residual holdup* led to a significant material accountancy failure at the Plutonium Fuel Production Facility (PFPF), a MOX fuel fabrication facility at Tokaimura, Japan. Residual holdup is defined as material that remains behind after the in-process material is removed for measurement prior to the taking of a physical inventory. Residual holdup resulting from the adhesion of powders on process equipment and accumulation in cracks, corners, and pores can result in persistent MUFs that grow with time. Ultimately, these MUFs can only be resolved by dismantlement and careful cleaning of process equipment.

At PFPF, operators noticed an unusually severe residual holdup problem soon after the plant started up in 1988. As a result, the plant operator, PNC, in conjunction with safeguards experts at Los Alamos, designed an nondestructive analysis (NDA) system to measure residual holdup in-situ known as the Glovebox Assay System (GBAS). However, measurement biases contributed to an overall uncertainty of about 15 percent. By 1994, the plant MUF had grown to about 69 kilograms of plutonium. Even if this entire amount was residual holdup, given the measurement uncertainty associated with the GBAS, the IAEA could not exclude the possibility with a confidence level of 95 percent, based on NDA measurements alone, that at least 1 SQ had been diverted. Consequently, the IAEA wanted PNC to cut open the plant gloveboxes, remove the holdup directly, and measure it with destructive assay (DA) methods. PNC balked at this request, and the dispute remained unresolved until the Nuclear Control Institute (NCI) publicly disclosed the existence of the discrepancy in 1994, after which PNC agreed to shut down the plant, recover the holdup, and install new equipment to reduce further holdup accumulation and improved NDA systems for measuring residual holdup more accurately. After an expenditure of $100 million to remove and clean out old gloveboxes and install new ones, PNC announced in November 1996 that it had reduced the MUF to less than 10 kilograms (but not less than 1 SQ). This partial resolution of the MUF took more than two years after the situation became public.

Another long-unresolved MUF issue at Tokaimura was associated with the accumulation of plutonium-laden fuel scrap resulting from decades of MOX research and production activities at the site.[8] Press reports in the mid-1990s indicated that the scrap inventory at Tokaimura contained between 100 and 150 kilograms of plutonium.[9] However, much of this scrap was in an impure form that could not be accurately measured via NDA methods. An instrument known as the Plutonium Scrap Multiplicity Counter (PSMC), developed by Los Alamos, was good for assaying clean scrap but was much less useful for assaying plutonium that was contaminated with moisture or other substances containing light elements that could generate neutrons through (α,n) reactions. For such heavily contaminated scrap, the measurement precision ranged from 10- 50 percent, which is well over the 4 percent uncertainty cited by the IAEA as the acceptable international standard for scrap measurements.[10] At an average precision of 10 percent, the uncertainty associated with measuring a scrap inventory containing 150 kilograms of plutonium would be greater than 1 SQ, and the 95 percent confidence level for detecting diversion would be over six SQs. Consequently, the IAEA wanted the plant operator, PNC, to chemically purify the scrap so that it could be made homogeneous and could be more precisely measured using DA. PNC apparently had long-range plans to build a facility for aqueous processing of the scrap, but a formal agreement with the IAEA that it would do so was not reached until 1998, when the IAEA announced that the plant operator would embark on a 5-year program "aimed at reducing the inventory of heterogeneous scrap material," which would be "gradually homogenized to allow enhanced verification, including destructive analysis."[11] Aside from a short mention in the IAEA 2000 Safeguards Statement of the implementation of a containment and surveillance approach for the receipt and storage of MOX scrap at something called the "Critical Solution Facility" in Japan,[12] no public information could be located by the author regarding the status or outcome of this program.

Measurement and estimation errors also contributed to substantial material accountancy failures that occurred at the spent fuel reprocessing plant at Tokaimura since it began operating in 1977. In January 2003, Japan admitted that the cumulative shipper-receiver difference—that is, the difference between the amount of plutonium that was estimated to have been shipped to the reprocessing plant and the amount that had actually been measured—was 206 kilograms, or about 25 SQs. This was nearly 3 percent of the total amount of plutonium estimated to have been processed in the plant over its lifetime. A few months later, Japan revised its figures, claiming that the actual discrepancy was 59 kg, with the remainder either bound in cladding hulls (12 kg), discarded with high-level liquid waste (106 kg), or decayed into americium-241 (29 kg). However, it was unclear how figures as precise as these were derived, given the uncertainties inherent in measuring the plutonium in cladding hulls and in high-level waste, and the uncertainties in determining the initial isotopic composition.

MUF issues have arisen in other countries that produce and process plutonium, including France and the United Kingdom. The Euratom Safeguards Agency reported in 2002 that "the annual verification of the physical inventory of the Cogema-Cadarache plant in France found an unacceptable amount of material unaccounted for (MUF) on the plutonium materials."[13] The problem was later attributed to issues associated with differences between measurement results taken by inspectors and operators and with the accounting of poorly-defined historical materials (although it is unclear why the issue did not arise until 2002, if that were the case). The finding of "high values of MUF" was reiterated in the 2003 report. It was reported in September 2004 that Euratom had "recently" sent a response to Cogema accepting its explanation for the 2002 MUF finding. Thus it took at least 2 years to resolve the discrepancy (and the time period would potentially be much longer if it was due to long-stored historical materials).

In the United Kingdom (UK), the most recent audit of nuclear materials at BNFL Sellafield, published on February 17, 2005, revealed a plutonium MUF of 29.6 kilograms, or about 3.5 SQ. BNFL insisted at the time that the figure did not mean that any material had been removed from its plants, and that "the techniques we use to account for our nuclear material are internationally approved and recognized as best practice. In particular, the systems of statistical measurement and control in the Thermal Oxide Reprocessing Plant (THORP) estimate the amount of plutonium are the most advanced in the world. . . ."[14]

BNFL was forced to eat these words only a few months later when the public was informed on May 9 of a massive leak at THORP that had gone undetected for 9 months. The leak, which occurred in a feed pipe to one of the two accountancy vessels, resulted in the accumulation of 83.4 cubic meters of dissolver solution, containing an estimated 19 metric tons of uranium and 190 kilograms of plutonium.[15] In spite of the fact that the leak occurred at an accountancy tank, which is where the initial inventory is measured for the purpose of establishing shipper-receiver differences (SRDs), the steadily increasing loss of material did not attract notice until 8 months after it began. To the credit of the plant's material accounting system, first indications of a problem came not from any safety indicators (several of which were malfunctioning), but from the Safeguards Department, when it detected an anomalous SRD in March 2005. However, an unambiguous finding of a leak did not take place until a month later.

In BNFL's review of the incident, it commended the role of the Safeguards Department in detecting the leak, but pointed out that the Nuclear Materials Accountancy system "is intended to provide overall accountancy balances," but "is not designed to (nor is it intended that it should) be responsive to track material on a more real time basis." BNFL goes on to recommend introduction of "a nuclear material tracking regime . . . with the objective of promptly detecting primary containment failure or misdirection of material."[16]

This statement appears completely baffling in view of the claims that BNFL had made previously, and fully supported by Euratom, regarding the status of near real time accountancy at THORP. For instance, in a paper delivered at an IAEA safeguards symposium in 2001, a joint paper by BNFL and Euratom safeguards officials stated that "Near Real Time Materials Accountancy (NRTMA) is fully operational in THORP, providing regular assistance of high quality material control."[17] One can only conclude that this claim was a bluff—a bluff that has now been called.

Even more troubling than the control failures on the part of the operator was that Euratom also appeared to be asleep at the wheel. THORP is allegedly under Euratom safeguards, which is charged with verifying that there has been no diversion of plutonium, based on a timeliness criterion identical to that of the IAEA (one SQ within 1 month). In addition to having access to the operators' accountancy data, Euratom apparently also had independent access to process data, upon which it performed its own statistical tests.[18] Yet there is no indication that Euratom inspectors were any more successful than the plant operators at detecting the leak and sounding an alarm. If this was indeed the case, this incident does not instill confidence in the ability of Euratom safeguards to detect a diversion.

Have Things Improved?

While the above real-world examples demonstrate the practical difficulties of ensuring through material accountancy methods the timely detection of diversions of significant quantities of plutonium at large, complex, messy, bulk-handling facilities, it is reasonable to ask whether they are representative of the situation today. After all, these facilities by and large are fairly old, and were planned and built decades ago; many of the most challenging material accountancy problems resulted from processes that were not optimized for safeguards effectiveness or from inventories of poorly characterized legacy materials. Can't we do better now?

Miller observed that his assessment of the limitations of material accountancy could change if improvements were made in the technical capabilities of material accountancy tools. In particular, he cited (1) a reduction in the overall measurement uncertainty in the chemical process area; (2) the use of near-real-time accountancy on a weekly basis to improve the sensitivity of tests for protracted diversion; and (3) a reduction in the measurement error of plutonium in waste streams such as cladding hulls and sludges.

With regard to (1), perhaps the best indication that there has been little progress in reducing measurement uncertainties since Miller's paper was written is the fact that the IAEA "expected measurement uncertainty" associated with closing a material balance at a reprocessing plant remains 1 percent as of 2001, the same value reported by Miller in 1990.[19]

With regard to (2), near-real-time accountancy (NRTA) is a method in which inventories are taken and material balances closed on a much more frequent basis than the conventional annual physical inventory. By reducing the throughput of material associated with a material balance, the ability to detect diversions is improved. For instance, Miller showed that the threshold for detection of an abrupt diversion of 1 SQ of plutonium at a large bulk-handling plant could be accomplished by use of NRTA with physical inventories carried out on a weekly basis. However, given that the time to take a physical inventory of a large facility is approximately 1 week, including preparation time, cleanout of process equipment, measurement of the inventory, and reconciliation of anomalies,[20] such a high frequency of physical inventories is utterly impractical. Thus NRTA must utilize inventory measurements of in-process materials where possible, and its effectiveness will depend in large part on the uncertainties associated with these measurements. A major question is, therefore, whether NDA techniques have improved over the past 15 years to the extent that the benefits of NRTA can be fully realized. The uncomfortable fact of the leak at THORP, where NRTA was purportedly "fully operational," tends to raise doubts as to whether NRTA is yet capable of fulfilling its promise.

Finally, with regard to (3), considerable efforts at Los Alamos National Laboratory (LANL) and elsewhere have been made over the last decade to improve the capabilities of NDA instruments for waste measurements. The development of neutron multiplicity counters and high-efficiency epithermal neutron counters showed some promise in improving the precision of plutonium in waste drums. However, as was seen above, these instruments perform best when measuring well-characterized and pure materials, but provide marginal benefit when measuring low-assay, contaminated, and heterogeneous plutonium materials.

Any comprehensive assessment of the capabilities of material accountancy at large bulk-handling facilities today must include a review of the safeguards approach for the Rokkasho Reprocessing Plant (RRP), which is the only large-scale commercial reprocessing plant where IAEA safeguards are being applied. The safeguards system at Rokkasho, which has been under development since the early 1990s, is the product of a massive multinational effort and should be regarded as the state-of-the-art.

Independent of the technical capabilities of the safeguards system at RRP are two overarching points. First, according to members of the team who developed the safeguards approach, "the most important factor leading to the success" of meeting all the challenges of developing a safeguards system on the scale needed for the RRP is "the open and full cooperation between all parties—the IAEA, the State, and the operator."[21] Therefore, even the most fully developed and technically sophisticated safeguards system will likely fail in the context of an uncooperative or adversarial relationship between these parties, which is exactly the situation of most interest in considering the future of IAEA safeguards as an instrument for controlling the use of nuclear energy not only in friendly states but in potentially adversarial ones. Second, issues of cost and convenience played a major role in development of the safeguards approach and resulted in many questionable compromises. For instance, instead of having its own independent on-site analytical laboratory, the IAEA must share a laboratory with the facility operator. Clearly, this situation raises additional complications, such as the potential for tampering, that must be addressed.

There is insufficient information in the public domain of the safeguards approach at Rokkasho for this author to make an independent assessment. However, it is clear that even after 15 years of designing the safeguards approach, the IAEA itself admits that its detection goals cannot be met at the facility. According to Shirley Johnson, former head of the Rokkasho safeguards project in the IAEA's Department of Safeguards,[22]

> The overall measurement uncertainty [at the RRP] may be less than +/-1%. This we won't know until we get further into Active Commissioning. However, even if it is 0.7% or 0.8% the fact remains that we cannot achieve the IAEA goal of 1 SQ detection capabilities. This has always been known. It comes down to a fact of very large throughput ... It is why it has taken us 15 years to develop the SG [safeguards] approach ... we had to compensate for lack of detection capabilities by enhancing our assurance that the facility operations are as declared ... all major flows of nuclear material ... are continuously monitored ...

Ms. Johnson said earlier during a talk at the 47th Institute of Nuclear Materials Management Annual Meeting in 2006 that the measurement uncertainty at RRP remained at 80 kilograms a year (corresponding to 1% of throughput, the same assumed by Miller in 1990), and that higher sensitivity and reliability of measurements were needed to improve on this.

Recent results from the performance of NDA solution monitoring systems at RRP indicate that they themselves have high measurement uncertainty. For instance, it was reported that the Plutonium Inventory and Management System (PIMS), which is designed to perform assays on relatively pure plutonium and uranium mixtures, has a total measurement uncertainty of +/-6%.[23]

Conclusion.

The bottom line is that nuclear material bulk-handling facilities, like other industrial facilities, are messy affairs. Although society may tolerate small leaks from a chemical plant to the environment if the hazards are limited, when the material in question can be used to build nuclear weapons, there is no acceptable level of leakage into the hands of hostile states or terrorists. The consequences of a single nuclear weapon falling into the wrong hands would be so catastrophic that there must be a zero-tolerance policy for diversion. If this standard cannot be met, then the underlying basis for claims that the closed fuel cycle can be adequately safeguarded against malevolent uses must be called into question.

ENDNOTES - CHAPTER 5

1. Pierre Goldschmidt, "Nuclear Proliferation in the 21st Century: Will Multilateral Democracy Work?" Copenhagen: Danish Institute for International Studies (DIIS), August 25-26, 2005, p. 2-3.

2. Marvin M. Miller, "Are IAEA Safeguards on Plutonium Bulk-Handling Facilities Effective?" Washington, DC: Nuclear Control Institute, 1990; reprinted in *Nuclear Power and the Spread of Nuclear Weapons*, P. Leventhal et al., eds., Washington, DC: Brassey's, 2002, Appendix D, p. 273.

3. IAEA, "Safeguards Statement for 2004," para. 52, p. 11.

4. Miller.

5. U.S. Congress, Office of Technology Assessment, Nuclear Safeguards and the International Atomic Energy Agency, OT-ISS-615, Washington, DC: U.S. Government Printing Office, June 1995, p. 73.

6. Paul Leventhal, "Safeguards Shortcomings—A Critique," Washington, DC: Nuclear Control Institute, September 12, 1994; quoting the IAEA Safeguards Glossary, 1987 Ed., p. 27.

7. Office of Technology Assessment, 1995, p. 117. The IAEA Safeguards Criteria are not made publicly available, but presumably OTA staff had access to them.

8. Edwin S. Lyman, "Japan's Plutonium Fuel Production Facility: A Case Study of the Challenges of Nuclear Material Accountancy," 39th Annual Meeting of the Institute of Nuclear Materials Management (INMM), Naples, FL, July 1998.

9. Mark Hibbs, "PFPF Holdup Pu Inventory Under 10 Kg; R&D Work to Focus on Monju Fuel," NuclearFuel, November 4, 1996, p. 15.

10. *IAEA Safeguards Glossary*, 2001 Ed., Table III, p. 53.

11. *IAEA 1998 Annual Report*, p. 65.

12. IAEA 2000 Annual Report, p. 102, available from *www.iaea.org/Publications/Reports/Anrep2000/safeguards.pdf.*

13. European Commission, "Report from the Commission to the European Parliament and Council: Operation of Euratom Safeguards in 2002," p. 9.

14. BNFL, "Media Response: Publication of Materials Unaccounted For (MUF)," press release, February 17, 2005.

15. BNFL, Board of Inquiry Report, "Fractured Pipe With Loss of Primary Containment in the THORP Feed Clarification Cell," May 26, 2005, p. 5.

16. *Ibid.*, p. 15.

17. Barry Jones et al., BNFL, Claude Norman et al., European Commission, Euratom Safeguards Office, "NRTMA: Common Purpose, Complementary Approaches," IAEA-SM-367/8/03, IAEA Safeguards Symposium, October-November 2001, IAEA, Vienna, Austria.

18. *Ibid.*, p. 5.

19. *IAEA Safeguards Glossary*, 2001 Ed., p. 23.

20. Thomas G. Clark et al., Westinghouse Savannah River Company, "Continuous Material Balance Reconciliation for a Modern Plutonium Processing Facility," 45th Annual Meeting of the Institute of Nuclear Materials Management (INMM), July 18-22, 2004, Orlando, FL.

21. Shirley Johnson et al., "Meeting the Safeguards Challenges of a Commercial Reprocessing Plant," 7th International Conference on Facility Operator-Safeguards Interface," February 29-March 4, 2004, Charleston, SC.

22. Shirley Johnson, IAEA, personal communication, July 27, 2006.

23. Y. Noguchi, "Validation and Performance Test of Plutonium Inventory Measurement System (PIMS) at Rokkasho Reprocessing Plant (RRP), Institute of Nuclear Materials Management 48th Annual Meeting, Tucson, AZ, July 8-12, 2007.

CHAPTER 6

ADEQUACY OF IAEA'S SAFEGUARDS FOR ACHIEVING TIMELY DETECTION

Thomas B. Cochran

INTRODUCTION

The purpose of this chapter is to examine, in light of the A. Q. Kahn network in Pakistan and recent events in Iran and North Korea, the adequacy of the International Atomic Energy Agency's (IAEA) safeguards for achieving timely detection of an effort to acquire nuclear weapons by a non-weapon state. For those less familiar with the obligation of state members of the Treaty on the Non-Proliferation of Nuclear Weapons (Non-Proliferation Treaty, or NPT) and/or states that operate under agreements with the IAEA, the Appendix to this chapter includes relevant excerpts from the NPT, the IAEA's enabling statute, and other IAEA publications.

THE OBJECTIVE OF SAFEGUARDS

As set forth in Article III.1 of the NPT, a primary purpose of IAEA's safeguards system is to prevent "diversion of nuclear energy from peaceful uses to nuclear weapons or other nuclear explosive devices." (See Appendix, Non-Proliferation Treaty).

Since Article III.1 of the NPT stipulates that IAEA safeguards *shall be followed*, any violation of IAEA safeguards is a violation of Article III of the NPT and therefore a violation of the treaty. Thus, when observers point out that the IAEA has no mandate to verify compliance with the NPT but only compliance with IAEA safeguards agreements, this is at best misleading since failure to comply with an applicable IAEA safeguards agreement is a violation of the NPT.

As set forth in the IAEA's enabling statute, IAEA safeguards are "designed to ensure that special fissionable and other materials, services, equipment, facilities, and information made available by the Agency or at its request or under its supervision or control are not used in such a way as to further any military purpose . . ." (see Appendix, IAEA's Enabling Statute).

The IAEA's enabling statute gives the IAEA certain rights. Among them is the right to establish an inspection system that is designed to ensure that the purpose of the safeguards is met. IAEA document INFCIRC/153, which details the safeguards obligations of states that are party to the NPT, provides a technical definition of the object of IAEA safeguards, namely, "the objective of safeguards is the timely detection of diversion of significant quantities of *nuclear material* from peaceful activities to the manufacture of nuclear weapons or of other explosive devices or for purposes unknown, and deterrence of such diversion by the risk of early detection."[1]

KEY SAFEGUARDS TERMS

The key terms of the objective of safeguards were not defined in INFCIRC/153; this task was given to the Standing Advisory Group on Safeguards Implementation (SAGSI) of the IAEA, an advisory group of technical safeguards experts.[2]

SAGSI considered the problem of quantifying the safeguards objective for several years. It identified four terms appearing either explicitly or implicitly in the statement of the objective just quoted as in need of quantitative expression. These were: significant quantities, timely detection, risk of detection, and the probability of raising a false alarm. It defined the associated numerical parameters (significant quantity, detection time, detection probability, and false alarm probability) as detection goals.[3]

In 1977, SAGSI submitted numerical estimates for these goals to the Director of Safeguards of the IAEA. The values recommended by SAGSI for the detection goals were carefully described as provisional guidelines for inspection planning and for the evaluation of safeguards implementation, not as requirements, and were so accepted by the Agency.[4] They have since been incorporated in the *IAEA Safeguards Glossary*, excerpts of which are reproduced below and in the Appendix.

Significant Quantity.

Significant quantity (SQ) is the approximate amount of nuclear material for which the possibility of manufacturing a nuclear explosive device cannot be excluded. Significant quantities take into account unavoidable losses due to conversion and manufacturing processes and should not be confused with critical masses.[5] Significant quantity values currently in use by the IAEA are given in Table 1.

In a previous Natural Resources Defense Council (NRDC) report, we argued that the IAEA's SQ values for direct use materials are not technically valid or defensible, and it was proposed that the SQ values for direct use plutonium and HEU be reduced by a factor of about eight.[6] Table 2 gives the approximate plutonium

Material	SQ
Direct Use Nuclear Material	
Pu[a]	8kg Pu
^{233}U	8kg ^{223}U
Highly enriched uranium [HEU] (^{235}U>20%)	25kg ^{235}U
Indirect Use Nuclear Material	
U (^{235}U < 20%)[b]	75kg ^{235}U (or 20t natural U or 20 t depleted U)
Th	20 t Th
a. For Pu containing less than 80 percent ^{238}Pu. b. Including low enriched natural and depleted uranium.	

Table 1. Significant Quantities.[7]

WEAPON-GRADE PLUTONIUM (kg)				HIGHLY-ENRICHED URANIUM (kg)		
Yield	Technical Capability			Technical Capability		
(kt)	Low	Medium	High	Low	Medium	High
1	3	1.5	1	8	4	2.5
5	4	2.5	1.5	11	6	3.5
10	5	3	2	13	7	4
20	6	3.5	3	16	9	5
Values rounded to the nearest 0.5 kilogram.						

Table 2. NRDC Estimate of the Approximate Fissile Material Requirements for Pure Fission Nuclear Weapons.[8]

and HEU requirements for pure fission weapons as estimated by NRDC. Regarding indirect use material, we note that 375 kilograms (kg) of 20 percent-enriched uranium, which contains one SQ (75kg of ^{235}U), when enriched, using a tails assay of 0.2 to 0.3 percent, yields

79-80kg of 93.5 percent-enriched product, which is three times larger than the SQ for direct use HEU. While it is not the purpose of this chapter to reexamine the validity of the SQ values, we simply note the obvious: if the SQ values are substantially lowered, it could significantly impact estimated conversion times.

Detection Time.

Detection time is the maximum time that may elapse between diversion of a given amount of nuclear material and detection of that diversion by IAEA safeguards activities. Where there is no additional protocol in force or where the IAEA has not drawn a conclusion of the absence of undeclared nuclear material and activities in a state (see *IAEA Safeguards Glossary*, No. 12.25), it is assumed that: (a) all facilities needed to clandestinely convert the diverted material into components of a nuclear explosive device exist in a state; (b) processes have been tested (e.g., by manufacturing dummy components using appropriate surrogate materials); and (c) nonnuclear components of the device have been manufactured, assembled, and tested. Under these circumstances, **detection time should correspond approximately to estimated conversion times** (see *IAEA Safeguards Glossary*, No. 3.13). Longer detection times may be acceptable in a state where the IAEA has drawn and maintained a conclusion of the absence of undeclared nuclear material and activities. Detection time is one factor used to establish the timeliness component of the IAEA inspection goal (see *IAEA Safeguards Glossary*, No. 3.24).[9] [Emphasis added]

Conversion Time.

Conversion time is the time required to convert different forms of nuclear material to the metallic components of a nuclear explosive device. Conversion time does not include the time required to transport diverted material to the conversion facility or to assemble the device, or any subsequent period. The diversion activity is assumed to be part of a planned sequence of actions chosen to give a high probability of success in manufacturing one or more nuclear explosive devices with minimal risk of discovery until at least one such device is manufactured.[10] The conversion time estimates applicable at present under these assumptions are provided in Table 3.

Beginning Material Form	Conversion Time
Pu, HEW, or ^{233}U metal	Order of days (7-10)
PuO_2, $PU(NO_3)_4$ or other pure Pu compounds; HEU or ^{233}U oxide or other pure U compounds; MOX or other nonirradiated pure mixtures containing Pu, U (^{233}U+^{235}U>20%); Pu, HEU, and/or ^{233}U in scrap or other miscellaneous impure compounds	
PU, HEU, or ^{233}U in irradiated fuel	Order of months (1-3)
U containing <20% ^{235}U and ^{233}U; Th	Order of months (3-12)
a This range is not determined by any single factor, but the pure Pu and U compounds will tend to be at the lower end of the range and the mixtures and scrap at the higher end.	

Table 3. Estimated Material Conversion Times for Finished Pu or U Metal Components.[11]

IAEA Timeliness Detection Goal.

The IAEA timeliness detection goal is the target detection times applicable to specific nuclear material categories (see *IAEA Safeguards Glossary,* No. 4.24).

These goals are **used for establishing the frequency of inspections** (see No. 11.16) and safeguards activities at a facility or a location outside facilities during a calendar year to verify that no abrupt diversion (see *IAEA Safeguards Glossary,* No. 3.10) has occurred. Where there is no additional protocol in force or where the IAEA has not drawn and maintained a conclusion of the absence of undeclared nuclear material and activities in a state (see *IAEA Safeguards Glossary,* No. 12.25), **the detection goals are as follows:**

- **One month for unirradiated direct use material,**
- **Three months for irradiated direct use material,** and
- **One year for indirect use material**.

Longer timeliness detection goals may be applied in a state where the IAEA has drawn and maintained a conclusion of the absence of undeclared nuclear material and activities in that state.[12]

With regard to the IAEA's timeliness detection goals, it should be noted that the Agency's resource limitations and resistance of member countries keep the actual inspection frequencies lower than the goals.[13]

ADEQUACY OF CONVERSION TIMES AND DETECTION GOALS

We now turn to the issue of the adequacy of the IAEA's estimated conversion times set forth in Table 3 above, and the timeliness detection goals set forth in paragraph 3.20, of the *IAEA Safeguards Glossary*. We begin with unirradiated direct use material.

Unirradiated Direct Use Material in Metal Form.

As seen in Table 3, the IAEA estimates that it will take a state on the "order of days (7-10)" to manufacture finished nuclear weapon components from plutonium, HEU or ^{233}U metal, where it is assumed that: (a) all facilities needed to clandestinely convert the diverted material into components of a nuclear explosive device exist in a state; (b) processes have been tested (e.g., by manufacturing dummy components using appropriate surrogate materials); and (c) that non-nuclear components of the device have been manufactured, assembled, and tested.

This is not an unreasonable estimate based on the time it took the United States to fabricate finished HEU components for the *Little Boy* device dropped on Hiroshima, Japan, on August 6, 1945. Consistent with the IAEA assumption, the non-nuclear components of *Little Boy* were assembled and tested before the all of the HEU was produced.

The HEU metal was shipped from Oak Ridge, Tennessee, to Los Alamos, New Mexico, in batches over a period of about a year. In the 6-week period from June 16 to July 28, Oak Ridge produced about 22kg of HEU. This was shipped to Los Alamos in batches of a few kilograms each. We estimate that the cumulative Oak Ridge production of HEU through July 14, 1945, was about 67kg, only 3kg in excess of what went into *Little Boy*. Thus, allowing for some losses, it is unlikely that Oak Ridge had produced enough HEU for *Little Boy* before that date.

The shipments of HEU metal from Oak Ridge to Los Alamos by road and rail typically took about 2 days. The shipment of the last six HEU finished components departed in three cargo planes carrying two components each from Kirtland Field, Albuquerque, New Mexico, on the morning of July 26, and, after stopping in San Francisco, arrived at Tinian in the Mariana Islands, South Pacific, on July 28. Thus, allowing for 2 days to transport the HEU metal from Oak Ridge to Los Alamos, Los Alamos must have fabricated the last of the HEU components in 9 days or less.

Little Boy was a gun-assembly type weapon requiring more than one SQ of HEU. The IAEA assumptions are based on an SQ value of 25kg of HEU, which implies an implosion device that would require the casting and machining of only one or two components. Moreover, although it took a relatively long time to enrich the HEU for *Little Boy*, this longer HEU production period is not a factor to be considered here. In sum, if we are correct that it took 9 days or less for Los Alamos to fabricate a few HEU *Little Boy* components in 1945, then 7-10 days is also reasonable assumption for the time it would take today for a state to manufacture finished components for an implosion-type weapon from an SQ amount of HEU.

Although the estimated detection time for direct use material appears reasonable, what is puzzling is that the timeliness detection goal is much longer, namely 1 month according to paragraph 3.20 of the *IAEA Glossary* (reproduced above). Moreover, both the estimated detection time and the timeliness detection goal, in our view, are far too short to allow time for diplomatic pressure to prevent the non-weapon state from fabricating a weapon. In fact, there is insufficient time to the IAEA staff to develop its report to the Board of Governors of the IAEA and for the Board of Governors to report to the UN Security Council.

Unirradiated Direct Use Material in Chemical Compounds and Mixtures.

As seen in Table 3, the IAEA estimates that it will take a state on the "order of weeks (1-3)" to manufacture finished components from unirradiated mixed-oxide (MOX) fuel, or from other compounds or mixtures containing plutonium, HEU or ^{233}U, again assuming that: (a) all facilities needed to clandestinely convert the diverted material into components of a nuclear explosive device exist in a state; (b) processes have been tested (e.g., by manufacturing dummy components using appropriate surrogate materials); and (c) non-nuclear components of the device have been manufactured, assembled, and tested.

Certainly, the lower end of this range—that is, on the order of a week—is a reasonable estimate of the time required in that it assumes that the time to convert the compound to a metal does not add appreciably to the time estimated to convert the metal to a finished component shape.

The upper end of the range—on the order of 3 weeks—seems unnecessarily generous. For example, plutonium metal can be prepared by calcium reduction of plutonium fluorides or oxides in induction-heated MgO crucibles, under an inert atmosphere of helium or argon.[14] Preparation of plutonium metal by reduction of a halide with an alkali or alkaline earth metal in a sealed pressure bomb is the only facet of chemical processing of plutonium that has remained unchanged over the years.[15] Using this technique, a few SQs of plutonium could readily be prepared in a small hot cell in a few days' time.

In any case, whether starting with unirradiated direct use material in metal or compound form, setting a timeless detection goal of 1 month is longer than any conservative estimate of the conversion time and shorter than the time required to bring diplomatic pressure to bear to halt the program.

Pu, HEU or ^{233}U in Irradiated Fuel.

Here the IAEA estimates a conversion time of on the "order of months (1–3)" and the IAEA's timeliness detection goal is 3 months. Assuming the plutonium or HEU is in irradiated fuel, the state must reprocess the fuel, convert the product into metal, and fabricate finished components. With regard to the reprocessing step, there are three diversion cases to consider: 1) the state already operates one or more reprocessing plants, pilot plants, or hot cells under IAEA safeguards; 2) it possesses a clandestine pilot reprocessing plant or hot cell; or 3) the state constructs a small "quick and dirty" reprocessing plant. Another important consideration is the spent fuel cooling time, that is, the time period between the removal of the irradiated fuel from the reactor and commencement of reprocessing.

Due to the high radioactivity levels and high thermal heat output associated with high burnup spent fuel from power reactors, the irradiated fuel is cooled 180 days or longer prior to reprocessing. For low burnup fuel, e.g., fuel elements or target materials removed from plutonium production reactors, the irradiated fuel can be processed after a shorter cooling period. In the United States during the Manhattan Project, the first fuel elements removed from the Hanford production reactors in late-1944 and early-1945 were chemically processed after only about 32-50 days of cooling time. Plutonium product was removed within a week of initiation of the batch processing.

If a state such as Japan already operates a declared reprocessing plant under safeguards, it could divert limited quantities of separated plutonium from plant operations with a low probability of detection by the IAEA, absent an informer. The inventory difference (ID) of reprocessing plants is on the order of 0.5 to 1 percent of the fuel throughput. High burnup, light water reactor (LWR), spent fuel typically contains approximately 1 percent plutonium. Thus, a pilot-scale reprocessing plant, if it processed 80 tons (t) of LWR spent fuel per year, would have an annual cumulative ID of about 0.5 to one SQ of plutonium. Some large-scale commercial reprocessing plants have a capacity that is 10 times greater.

Thus, a state with a large declared reprocessing plant under IAEA safeguards could divert an SQ of plutonium without detection over a period of about 1 month. A state with a pilot-size plant could divert the same quantity over a period of 1 year.

Some advanced reprocessing technologies contemplate not completely separating the plutonium from some actinides and fission products. While this should make it more difficult for an insider to divert plutonium, it would not represent a significant added barrier to a state effort to divert plutonium. Given that the added actinides and fission products would not add significantly to the plutonium mass, the state could divert the spiked plutonium to a small clandestine hot cell for additional processing. The processing time to recover an SQ of plutonium should take only a few days.

If a state does not have an existing declared reprocessing facility, it has the option of developing a clandestine capability, such as the Israeli facility hidden for years below the Dimona reactor. Alternatively, the state could attempt to develop a "quick and dirty" reprocessing capability. The feasibility of clandestine reprocessing of LWR fuel has been addressed by Oak Ridge National Laboratory,[16] Sandia Laboratories,[17] and others, and these studies have been reviewed by Marvin Miller.[18]

"The [Oak Ridge] study concluded the [reprocessing] plant could be in operation 4 to 6 months from the start of construction, with the first 10 kilograms of plutonium metal (about two bomb's worth) produced about 1 week after start of operation. Once in operation, the small plant could process about one PWR [pressurized water reactor] assembly per day, which translates into production of about 5 kilograms of plutonium per day."[19]

The 1966 Sandia study estimated the preparation lead-time for producing the first kilograms of plutonium employing a staff of six technicians was about 8 months.[20]

In sum, if a state has a declared pilot-scale or larger reprocessing plant, the conversion time should be the same as for unirradiated compounds of direct use materials, since the state could divert unirradiated compounds of direct use materials without being detected by the IAEA.

Low Enriched Uranium.

Here the IAEA estimates a conversion time of on the "order of months (3-12)" and the IAEA's timeliness detection goal is 1 year. The enrichment work, measured in kilograms of separative work units (kg SWU, often abbreviated SWU), required to obtain one SQ of HEU is a function of ^{235}U concentration of the uranium feed, product, and tails. Marvin Miller has identified and reviewed the major proliferation risks associated with centrifuge enrichment plants: (1) secret use of a declared, safeguarded low-enriched uranium (LEU) plant to produce HEU or exceeds LEU covertly; (2) construction and operation of a clandestine plant to produce HEU; and (3) conversion of a declared, safeguarded LEU plant to HEU production following breakout.[21]

According to Miller:

(1) The basic "Hexapartite" safeguards approach for centrifuge plants was developed during the early 1980s by a group of six countries—Germany, the United Kingdom, and the Netherlands (the URENCO states), and the United States, Japan, and Australia. It consists of two sets of activities:

(a) verifying the uranium material balance by measuring the amount of uranium as UF6 introduced into the plant as feed material and withdrawn as enriched product and tails; and.

(b) verifying that no material beyond the declared enrichment level, in particular, no HEU is being produced.

While (a) doesn't require inspector access to the cascade halls where the centrifuges are installed, (b) does, and the inspection procedures were designed to provide an element of surprise in order to deter production of HEU between routinely scheduled inspections, while also accounting for the plant operator's concern about the inspector's gaining knowledge of proprietary information relating to the construction and operation of the centrifuges. Various technical difficulties have been encountered over the years in applying (b) at specific plants. But confidence in the IAEA's ability to detect illicit production of HEU has improved dramatically since 1995 with the introduction of sampling and subsequent analysis of particles deposited on surfaces in the cascade area as a standard safeguards tool. Since release of particles to the plant environment is difficult to avoid and the analysis is highly precise, environmental sampling has emerged as a significant deterrent to clandestine HEU production in a declared LEU plant. On the other hand, current safeguards procedures cannot detect the production of LEU in excess of what the plant operator declares to be the normal production rate,[22] and this can significantly increase the difficulty of detecting a clandestine plant, as we discuss next.

(2) The much smaller energy consumption and process area characteristic of centrifuge plants compared to gaseous diffusion plants of the same separative capacity make the former much more difficult to detect. For example, a centrifuge plant with a separative capacity of 5,000 SWU/yr—sufficient to produce 25kg/yr of 90 percent enriched uranium—would likely require less than 100kW of power and have a "footprint" of about 500m^2.[23] Moreover, detection by wide area environmental monitoring is also difficult because emissions from a centrifuge plant normally are very small. The plant operates under high-vacuum conditions so that leaks primarily lead to an inflow of air into the centrifuge equipment, not to a significant release of UF6 from the system into the environment. Finally, as noted above, if excess LEU is used as feed for the clandestine plant instead of natural uranium,

the size of plant required to produce a given amount of HEU product is reduced significantly, especially if the tails concentration is also increased.

(3) There is the possibility of breakout, i.e., takeover by a state of a declared, safeguarded LEU centrifuge plant, and reconfiguration of the plant to produce weapons grade uranium.[24] Because of its high separation factor compared to the gaseous diffusion process, the inventory of a centrifuge plant is much smaller than a diffusion plant, and so is the equilibrium time, i.e., the time required to achieve full production after plant startup or subsequent modification, e.g., from production of LEU to production of HEU by recycling the product material back as feed. Typically, the equilibrium time for LEU centrifuge and diffusion plants are on the order of hours and months, respectively.

As noted by Gilinsky *et al.*, the SWU requirements to obtain one SQ of HEU can be reduced substantially if a state already has access to, and can successfully divert fresh LWR fuel.[25] In the examples given in Table 4, using 4 percent-enriched feed (typical of LWR fresh fuel) and operating the enrichment plant at a high tails assay—for example 2 percent ^{235}U—the separative work requirements are reduced by more than 80 percent of that required if natural uranium feed (0.711% ^{235}U) were used.

The enrichment plant capacity (SWU/y) is a product of the number of stages and the capacity of each stage. For a centrifuge enrichment plant, the capacity of a single stage is a function of length of the rotor and its peripheral speed.[26] In Table 4, we also calculate the number of centrifuge stages required to

Product (% ^{235}U)	93.5	93.5	93.5	93.5
Feed (% ^{235}U)	0.711	0.711	4.0	4.0
Tails (% ^{235}U)	0.25	0.5	0.25	2.0
Enrichment Work (kg SWU)	5,422	4,021	1,769	894
U Feed (tons) 1.144	5,057	11.02	0.622	
Centrifuges Required to Obtain 1 SQ/y[27]				
2 kg SWU/y/centrifuge (P1)	2,711	2,011	885	447
5 kg SWU/y/centrifuge (P2)	1,084	804	354	179
10 kg SWU/y/centrifuge (Russia)	542	402	177	89
40 kg SWU/y/centrifuge (URENCO)	136	101	44	22
300kg SWU/y/centrifuge (U.S. R & D)	18	13	6	3

Table 4. Enrichment Requirements to Obtain One SQ of HEU.

obtain one SQ per year of 93.5 percent-enriched HEU. As seen from Table 4, depending primarily on the feed enrichment and the efficiency of each stage, the number of centrifuge stages required to obtain one SQ if HEU per year varies from a few to a few thousand.

We know from events in Iran (and North Korea), a small centrifuge enrichment plant with up to a few hundred centrifuge stages can be readily hidden from the IAEA and from foreign intelligence efforts. A state can acquire the necessary technology and construct and operate a small clandestine centrifuge plant with little risk of detection, and the probability of detection is substantially reduced if the state has a declared centrifuge plant under safeguards.

Assuming a state may have a small clandestine enrichment plant, the conversion time could be on the order of weeks to months, depending on the number of size of the plant and the technology employed.

CONCLUSIONS

IAEA safeguards are inadequate for achieving the objective of timely detection of diversion of significant quantities of *nuclear material* from peaceful activities to the manufacture of nuclear weapons.

The IAEA's SQ values are technically erroneous and excessive.

For unirradiated direct use material in metal form, the IAEA's estimated conversion time (7-10 days) is adequate, but the timeliness detection goal (1 month) is too long, and timely warning cannot be achieved. Nonweapon states should not be permitted to possess an SQ of unirradiated direct use material in metal form.

For unirradiated direct use material in chemical compounds and mixtures, the IAEA's estimated conversion time is on the order of weeks (1-3). The lower end of this range is adequate, but the upper end appears too generous. The timeliness detection goal (1 month) is too long, and the timely detection cannot be achieved. Non-weapon states should not be permitted to possess an SQ of unirradiated direct use material in the form of chemical compounds or mixtures.

For plutonium, HEU or ^{233}U in irradiated fuel, the IAEA's estimated conversion time (1-3 months) is adequate. However, if a state possesses a safeguarded pilot-size or larger reprocessing plant, a state can divert SQs of separated plutonium from plant operations with a low probability of detection by the IAEA absent an informer. If a state has a declared pilot-scale or larger reprocessing plant, the conversion time should be the same as for unirradiated compounds of direct use materials.

Non-weapon states should not be permitted to possess pilot-scale or larger reprocessing plants. When conducted in non-weapon states, research on reprocessing and transmutation related technologies, including those that are unlikely to ever be commercialized, simply train cadres of experts in actinide chemistry and plutonium metallurgy, a proliferation concern in its own right. The hot cells, used for on-hands research, provide readily available facilities for separation of plutonium and fabrication of plutonium components for weapons. Thus, smaller reprocessing activities, and research and development on transmutation related technologies, should not be permitted in non-weapon states.

For indirect use material, such as low-enriched uranium, the IAEA's estimated conversion time is on the order of months (3-12). The lower end of this range is adequate, but the upper end appears too generous. Small gas centrifuge plants can be readily hidden from IAEA inspectors and foreign intelligence forces. If a state is permitted to possess a safeguarded enrichment plant, it can be used as a cover for procuring components and materials needed for a small clandestine plant. A state possessing a safeguarded centrifuge enrichment plant can rapidly reconfigure the plant to produce HEU. Also, a state may have a small clandestine enrichment plant. In either case, the conversion time could be on the order of weeks to months, depending on the number of and size of the plants and the technology employed.

Even if the IAEA's timeliness detection goal of 1 year is met, this is unlikely to provide "timely warning." Consequently, enrichment plants should not be permitted in non-weapon states.

In sum, our recommended conversion times are given in Table 5. The detection goals should be the lower end of the conversion time range in each case.

Beginning Material Form	Conversion Time
Pu, HEW, or ^{233}U metal	Order of days (7-10)
PuO_2 $PU(NO_3)_4$ or other pure Pu compounds; HEU or ^{233}U oxide or other pure U compounds; MOX or other nonirradiated pure mixtures containing PU, U (^{233}U+^{235}U>20%); Pu, HEU, and/or ^{233}U in scrap or other miscellaneous impure compounds	Order of days (7-10)
PU, HEU, or ^{233}U in irradiated fuel State without declared reprocessing Non-weapon states are no permitted to possess reprocesing plants	Order of months (1-3)
U containing <20% ^{235}U and ^{233}U; Th State without declared enrichment Non-weapon states are not permitted to posses enrichment plants	Order of weeks to months

Table 5. Recommended Material Conversion Times for Finished Pu or U Metal Components.

ENDNOTES - CHAPTER 6

1. International Atomic Energy Agency (IAEA), INFCIRC/153, Paragraph 28, Vienna, Austria: IAEA.

2. Marvin Miller, "Are IAEA Safeguards on Plutonium Bulk-Handling Facilities Effective?" Nuclear Control Institute, available from *www.nci.org/k-m/mmsgrds.htm*.

3. *Ibid*.

4. *Ibid*.

5. IAEA, *IAEA Safeguards Glossary*, 2001 Ed., International Verification Series, No. 3, 2002, Paragraph 3.14.

6. Thomas B. Cochran and Christopher E. Paine, "The Amount of Plutonium and Highly-Enriched Uranium Needed for Pure Fission Nuclear Weapons," Natural Resources Defense Council, revised April 13, 1995, Paragraph 3.13.

7. Table 1 is identified as Table II in the *IAEA Safeguards Glossary*.

8. *Ibid.*, p. 9.

9. *IAEA Safeguards Glossary*, 2001 Ed., Paragraph 3.15.

10. *Ibid.*, Paragraph 3.13.

11. *Ibid.* Table 3 here is identified as Table I in the glossary.

12. *Ibid.*, Paragraph 3.20.

13. Victor Gilinsky, Marvin Miller, and Harmon Hubbard, *A Fresh Examination of the Proliferation Dangers of Light Water Reactors*, Washington, DC: The Nonproliferation Policy Education Center, September 2004, p. 22, makes this point with regard to light water reactor (LWR) inspections.

14. Manson Benedict, Thomas Pigford, and Hans Levi, *Nuclear Chemical Engineering*, New York: McGraw-Hill Book Company, New York, 1981, p. 430.

15. *Plutonium Handbook*, Vol. I, O. J. Wick, ed., La Grange, IL: The American Nuclear Society, 1980, p. 564.

16. D. E. Ferguson to F. L. Culler, Intra-Laboratory Correspondence, "Simple, Quick Processing Plant," Oak Ridge, TN: Oak Ridge National Laboratory, August 30, 1972, 22 pp.

17. J. P. Hinton *et al.*, *Proliferation Resistance of Fissile Material Disposition Program(FMDP) Plutonium and Disposition Alternatives: Report of the Proliferation Vulnerability Red Team, Sandia National Laboratories*, Report No. SAND97-8201, October 1996, Section 4.1.1.3.

18. Marvin Miller, "The Feasibility of Clandestine Reprocessing of LWR Spent Fuel," Appendix 2, in Gilinsky, Miller, and Hubbard.

19. Gilinsky, Miller, and Hubbard, p. 21.

20. *Ibid.*, p. 23.

21. Marvin Miller, "The Gas Centrifuge and Nuclear Proliferation," Appendix. 1, p. 38, in Gilinsky, Miller, and Hubbard.

22. That is, the IAEA currently cannot verify the separative capacity of a centrifuge plant as stated by the operator. Thus, the operator could understate the plant's true separative capacity and feed undeclared uranium to the cascades, producing excess, undeclared LEU after the inspectors have left the plant following their monthly visits, which normally last several days.

23. This is based on the use of 5kg SWU/yr P2 centrifuges which each occupy an area of about 0.25 m^2 and have an energy consumption of about 150 kW/kg SWU.

24. Such reconfiguration can be accomplished in various ways depending on the plant design. For example, in URENCO plants, which consist of many parallel independent cascades each producing LEU product, the LEU product of one cascade can be used as feed material for another cascade, and so on, until the desired HEU product concentration is achieved. By contrast, centrifuge plants of Russian design are configured as one large cascade whose product and tails concentrations can be changed remotely from the plant control room by changing the valve connections on the centrifuges.

25. Gilinsky, Miller, and Hubbard.

26. Capacity scales as V_2L, where V is the peripheral speed and L is the rotor length.

27. Individual centrifuge capacity values are from Miller, "The Gas Centrifuge and Nuclear Proliferation," Appendix. 1, Table 1, in Gilinsky, Miller, and Hubbard.

APPENDIX

NON-PROLIFERATION TREATY

The Non-Proliferation Treaty (NPT) was signed July 1, 1968, and entered into force March 5, 1970. All non-weapon state parties to the NPT are required to comply with International Atomic Energy Agency (IAEA) safeguards, as indicated under Article III of the NPT Treaty:

> III.1. **Each non-nuclear-weapon State Party to the Treaty undertakes to accept safeguards,** as set forth in an agreement to be negotiated and concluded with the International Atomic Energy Agency **in accordance with the Statute of the International Atomic Energy Agency and the Agency's safeguards system,** for the exclusive purpose of verification of the fulfillment of its obligations assumed under this Treaty with a view to preventing diversion of nuclear energy from peaceful uses to nuclear weapons or other nuclear explosive devices. **Procedures for the safeguards required by this article shall be followed with respect to source or special fissionable material whether it is being produced, processed or used in any principal nuclear facility or is outside any such facility.** The safeguards required by this article shall be applied to all source or special fissionable material in all peaceful nuclear activities within the territory of such State, under its jurisdiction, or carried out under its control anywhere. (Emphasis added)

The following nuclear weapon states are **NOT** parties to the NPT:

- Israel
- Pakistan
- India
- Democratic People's Republic of Korea (DPRK or North Korea).

The first three, Israel, India, and Pakistan, are known to have nuclear weapons and have never been signatories to the NPT. The DPRK is believed to have nuclear weapons and has declared that it has possesses nuclear weapons. On January 10, 2003, DPRK announced that it was withdrawing from the NPT effective immediately. All other states of any consequence are members of the NPT, and with the exception of the United States, United Kingdom, France, Russia, and China, all are non-weapon states subject to IAEA safeguards.

IAEA'S ENABLING STATUTE

The IAEA was established in 1957, 11 years prior to the inception of the NPT. Under Article III, paragraph A. 5, of its enabling statute, the Agency is authorized:

> **To establish and administer safeguards designed to ensure that special fissionable and other materials, services, equipment, facilities, and information made available by the Agency or at its request or under its supervision or control are not used in such a way as to further any military purpose**; and to apply safeguards, at the request of the parties, to any bilateral or multilateral arrangement,

> or at the request of a State, to any of that State's activities in the field of atomic energy; (Emphasis added)

The safeguards system is defined primarily in Article XII of the IAEA Statute. Article XII of the IAEA Statute states in part:

> A. With respect to any Agency project, or other arrangement where the Agency is requested by the parties concerned to apply safeguards, **the Agency shall have the following rights and responsibilities** to the extent relevant to the project or arrangement:
>
> 1. **To examine the design of specialized equipment and facilities**, including nuclear reactors, **and to approve it only from the viewpoint of assuring that it will not further any military purpose**, that it complies with applicable health and safety standards, **and that it will permit effective application of the safeguards provided for in this article**;
>
> . . .
>
> 5. **To approve the means to be used for the chemical processing of irradiated materials solely to ensure that this chemical processing will not lend itself to diversion of materials for military purposes** and will comply with applicable health and safety standards; to require that special fissionable materials recovered or produced as a by-product be used for peaceful purposes under continuing Agency safeguards for research or in reactors, existing or under construction, specified by the member or members concerned; and **to require deposit**

with the Agency of any excess of any special fissionable materials recovered or produced as a by-product over what is needed for the above-stated uses in order to prevent stockpiling of these materials, provided that thereafter at the request of the member or members concerned special fissionable materials so deposited with the Agency shall be returned promptly to the member or members concerned for use under the same provisions as stated above.

6. To send into the territory of the recipient State or States **inspectors**, designated by the Agency after consultation with the State or States concerned, **who shall have access at all times to all places and data and to any person who by reason of his occupation deals with materials, equipment, or facilities which are required by this Statute to be safeguarded, as necessary to account for source and special fissionable materials supplied and fissionable products and to determine whether there is compliance with the undertaking against use in furtherance of any military purpose** referred to in sub-paragraph F-4 of article Xl, with the health and safety measures referred to in sub-paragraph A-2 of this article, and with any other conditions prescribed in the agreement between the Agency and the State or States concerned. Inspectors designated by the Agency shall be accompanied by representatives of the authorities of the State concerned, if that State so requests, provided that the inspectors shall not thereby be delayed or otherwise impeded in the exercise of their functions; (Emphasis added)

IAEA AGREEMENTS WITH MEMBER STATES

The IAEA administers its safeguards requirements pursuant to agreements that the IAEA has with member states.

As of November 2004, there were 138 member states and 65 intergovernmental and nongovernmental organizations worldwide having formal agreements with the Agency, and 232 safeguards agreements in force in 148 states (and with Taiwan) involving 2,363 safeguards inspections performed in 2003.[1] The DPRK joined the IAEA in 1974, but withdrew its membership on June 13, 1994; and Cambodia, which joined the IAEA in 1958, withdrew its membership on March 26, 2003.

Since the IAEA was established in 1957, over the years the IAEA safeguards requirements have been upgraded and strengthened. The more explicit requirements are set forth in a series of IAEA Information Circulars, the most important of which are INFCIRC/26 (the Agency's Safeguards approved by the Board of Governors on January 31, 1961), INFCIRC/66 (designed to be applied in any state that concluded a safeguards agreement), and INFCIRC/153 (used as a basis for agreements with states that are parties to the NPT-and the Additional Protocol.

The "Basic Undertaking" of IAEA safeguards agreements with other parties is currently set forth in INFCIRC/153 (Corrected), June 1972:

> **The Agreement should contain, in accordance with Article III.l of the Treaty on the Non-Proliferation of Nuclear Weapons), an undertaking by the State to accept safeguards, in accordance with the terms of the Agreement,**

on all source or special fissionable material in all peaceful nuclear activities within its territory, under its jurisdiction or carried out under its control anywhere, **for the exclusive purpose of verifying that such material is not diverted to nuclear weapons or other nuclear explosive devices**. (Emphasis added)

COMPREHENSIVE, OR FULL-SCOPE, SAFEGUARDS

A comprehensive safeguards agreement is an IAEA safeguards agreement that applies safeguards on all nuclear material in all nuclear activities in a state. These are primarily safeguards agreements pursuant to the NPT, concluded between the IAEA and non-nuclear-weapon state (NNWS) parties as required by Article III.1 of the NPT, but they also include agreements pursuant to the Tlatelolco Treaty; the *sui generic* agreement between Albania and the IAEA; and the quadripartite safeguards agreement between Argentina, Brazil, the Brazil-Argentine Agency for Accounting and Control of Nuclear Materials (ABACC), and the IAEA. As of July 19, 2005, 37 NNWS parties to the NPT have not yet brought into force comprehensive safeguards agreements with the IAEA. Most of these 37 countries do not have significant nuclear facilities.[2]

IAEA INFORMATION CIRCULAR 153

INFCIRC/153 places constraints on the agency's safeguards implementation:

IMPLEMENTATION OF SAFEGUARDS

The Agreement should provide that safeguards **shall be implemented** in a manner designed:

a. **To avoid hampering the economic and technological development** of the state or international cooperation in the field of peaceful nuclear activities, including international exchange *nuclear material* 2);

b. **To avoid undue interference in the state's peaceful nuclear activities**, and in particular in the operation of *facilities*; and

c. To be **consistent with** prudent management **practices required for the economic and safe conduct of nuclear activities**.

INFCIRC/153 defines the:

OBJECTIVE OF SAFEGUARDS

28. The Agreement should provide that the **objective of safeguards is the timely detection of diversion of significant quantities of *nuclear material* from peaceful nuclear activities to the manufacture of nuclear weapons** or of other nuclear explosive devices or for purposes unknown, and deterrence of such diversion by the risk of early detection.

29. To this end the Agreement should provide for the use of material accountancy as a safeguards measure of fundamental importance, with containment and surveillance as important complementary measures.

30. The Agreement should provide that the technical conclusion of the Agency's verification

activities shall be a statement, in respect of each *material balance area,* of the amount of material unaccounted for over a specific period, giving the limits of accuracy of the amounts stated.

Also, INFCIRC/153 calls for a:

NATIONAL SYSTEM OF ACCOUNTING FOR AND CONTROL OF NUCLEAR MATERIAL

31. The Agreement should provide that **the state shall establish and maintain a system of accounting for and control of all nuclear material** subject to safeguards under the Agreement, and that such safeguards shall be applied in such a manner as to enable the Agency to verify, in ascertaining that there has been no diversion of nuclear material from peaceful uses to nuclear weapons or other nuclear explosive devices, findings of the state's system. The Agency s verification shall include, *inter alia,* independent measurements and observations conducted by the Agency in accordance with the procedures specified in Part II below. The Agency, in its verification, shall take due account of the technical effectiveness of the state's system.

IAEA SAFEGUARDS GLOSSARY

The *IAEA Safeguards Glossary* includes the definitions of several terms that are important to a discussion of the adequacy of the IAEA's safeguards with respect to timely warning, namely "diversion rate," "conversion time," "significant quantity," and "detection time":[3]

3.10. Diversion rate—the amount of nuclear material which could be diverted in a given unit of time. If the amount diverted is 1 SQ or more (see No. 3.14) of nuclear material in a short time (i.e., within a period that is less than the material balance period [see No. 6.47]), it is referred to as an **"abrupt" diversion**. If the diversion of 1 SQ or more occurs gradually over a material balance period, with only small amounts removed at any one time, it is referred to as a **"protracted" diversion**.

3.13. Conversion time—the time required to convert different forms of nuclear material to the metallic components of a nuclear explosive device. Conversion time does not include the time required to transport diverted material to the conversion facility, or to assemble the device, or any subsequent period. The diversion activity is assumed to be part of a planned sequence of actions chosen to give a high probability of success in manufacturing one or more nuclear explosive devices with minimal risk of discovery until at least one such device is manufactured. The conversion time estimates applicable at present under these assumptions are provided in Table I. [Reproduced as Table 3 above.]

3.14. Significant quantity (SQ)—the approximate amount of nuclear material for which the possibility of manufacturing a nuclear explosive device cannot be excluded. Significant quantities take into account unavoidable losses due to conversion and manufacturing processes and should not be confused with critical masses. Significant quantities are used in establishing the quantity component of the IAEA inspection goal (see No. 3.23). Significant quantity values currently in use are given in Table II. [Reproduced as Table 1 above.]

3.15. Detection time—the maximum time that may elapse between diversion of a given amount of nuclear material and detection of that diversion by IAEA safeguards activities. Where there is no additional protocol in force or where the IAEA has not drawn a conclusion of the absence of undeclared nuclear material and activities in a state (see No. 12.25), it is assumed: (a) that all facilities needed to clandestinely convert the diverted material into components of a nuclear explosive device exist in a state; (b) that processes have been tested (e.g., by manufacturing dummy components using appropriate surrogate materials); and (c) that nonnuclear components of the device have been manufactured, assembled and tested. Under these circumstances, **detection time should correspond approximately to estimated conversion times** (see No. 3.13). Longer detection times may be acceptable in a state where the IAEA has drawn and maintained a conclusion of the absence of undeclared nuclear material and activities. Detection time is one factor used to establish the timeliness component of the IAEA inspection goal (see No. 3.24).

3.20. IAEA timeliness detection goal—the target detection times applicable to specific nuclear material categories (see No. 4.24).

These goals are **used for establishing the frequency of inspections** (see No. 11.16) and safeguards activities at a facility or a location outside facilities during a calendar year, in order to verify that no abrupt diversion (see No. 3.10) has occurred. Where there is no additional protocol in force or where the IAEA has not drawn and maintained a conclusion of the absence of undeclared nuclear material and activities in a state (see No. 12.25), **the detection goals are as follows:**

– One month for unirradiated direct use material,

– Three months for irradiated direct use material,

– One year for indirect use material.

Longer timeliness detection goals may be applied in a state where the IAEA has drawn and maintained a conclusion of the absence of undeclared nuclear material and activities in that state.

3.22. IAEA inspection goal – performance targets specified for IAEA verification activities at a given facility as required to implement the facility safeguards approach (see No. 3.3). **The inspection goal for a facility consists of a quantity component** (see No. 3.23) **and a timeliness component** (see No. 3.24). These components are regarded as fully attained if all the Safeguards Criteria (see No. 3.21) relevant to the material types (see No. 4.23) and material categories (see No. 4.24) present at the facility have been satisfied, and all anomalies involving 1 SQ or more of nuclear material have been resolved in a timely manner (see No. 3.26). (See also Nos 12.23 and 12.25.)

3.23. Quantity component of the IAEA inspection goal—relates to the scope of the inspection activities at a facility that are necessary for the IAEA to be able to draw the conclusion that there has been no diversion of 1 SQ or more of nuclear material over a material balance period and that there has been no undeclared production or separation of direct use material at the facility over that period.

3.24. Timeliness component of the IAEA inspection goal—relates to the periodic activities that are necessary for the IAEA to be able to draw the conclusion that there has been no abrupt diversion (see No. 3.10) of 1 SQ or more at a facility during a calendar year. (Emphasis added)

ADDITIONAL PROTOCOL

The Additional Protocol is a legal document granting the IAEA complementary inspection authority to that provided in underlying safeguards agreements. A principal aim is to enable the IAEA inspectorate to provide assurance about both declared and possible undeclared activities. Under the Protocol, the IAEA is granted expanded rights of access to information and sites, as well as additional authority to use the most advanced technologies during the verification process.[4]

At the end of the Persian Gulf War, the world learned about the extent of Iraq's clandestine pursuit of an advanced program to develop nuclear weapons. The international community recognized that the Agency's international inspection system needed to be strengthened in order to increase its capability to detect secret nuclear programs. After 4 years of work by the Secretariat of the Agency, an Agency committee agreed on a Model Additional Protocol (the "Model Protocol") for strengthening nuclear safeguards. The Model Protocol was approved by the Agency's Board of Governors in 1997. The Model Protocol was designed to be used to amend existing safeguards agreements to strengthen such safeguards by requiring NNWS to provide, *inter alia,* broader declarations to the Agency about their nuclear programs and nuclear-related activities, and by expanding the access rights of the Agency. The new safeguards measures become effective in each state when it brings its protocol into force.[5]

The Model Protocol requires states to report a range of information to the Agency about their nuclear and nuclear-related activities and about the planned developments in their nuclear fuel cycles. This includes expanded information about their holdings of uranium and thorium ores and ore concentrates and of other plutonium and uranium materials not currently subject to Agency safeguards, general information about their manufacturing of equipment for enriching uranium or producing plutonium, general information about their nuclear fuel cycle-related research and development activities not involving nuclear material, and their import and export of nuclear material and equipment.[6]

As of July 19, 2005, 69 states and Euratom have ratified Additional Protocols.[7] Thirty-three additional states have signed, but not ratified Additional Protocols, bringing the total number of states that have signed to 102. The IAEA Board has approved Additional protocols for six additional states that have not signed. Notable countries that have not signed an Additional Protocol include:

- Algeria (IAEA Board Approval)
- Argentina
- Belarus
- Brazil
- DPRK
- Egypt
- India
- Israel
- Pakistan
- Serbia and Montenegro
- Syria
- Thailand
- Venezuela
- Vietnam

ENDNOTES - CHAPTER 6 APPENDIX

1. Available from *www.iaea.org/About/by_the_numbers.html.*

2. The Republic of the Congo has two small research reactors, at least one of which is not operable, and Niger is involved in uranium mining.

3. *IAEA Safeguards Glossary*, 2001 Ed., International verification Series, No. 3, Vienna, Austria: International Atomic Energy Agency (IAEA), 2002.

4. Available from *www.iaea.org/Publications/Factsheets/English/sg_overview.html.*

5. Available from *www.state.gov/t/np/trty/11757.htm.*

6. Available from *www.state.gov/t/np/trty/11757.htm.*

7. The IAEA also applies safeguards, including the measures foreseen in the Model Additional protocol, in Taiwan.

CHAPTER 7

MANAGING SPENT FUEL IN THE UNITED STATES: THE ILLOGIC OF REPROCESSING

Frank von Hippel

I. SUMMARY

Since 1982, it has been U.S. policy, for nonproliferation and cost reasons, not to reprocess spent power-reactor fuel. Instead, the U.S. Department of Energy (DoE) is to take spent power reactor fuel from U.S. nuclear utilities and place it in an underground federal geological repository. The first U.S. repository is being developed under Yucca Mountain, Nevada. Originally, it was expected to begin taking fuel in 1998. However, project management problems and determined opposition by the State of Nevada are expected to delay its opening for at least 2 decades.

U.S. nuclear utilities, therefore, have been pressing the DoE to establish one or more centralized interim storage facilities for their accumulating spent fuel. They insist that a "nuclear renaissance," i.e., investments in new nuclear power plants, will not take place in the United States until the federal government demonstrates that it is able to remove the spent fuel from the reactor sites. U.S. state governments resist hosting interim spent fuel storage, however, out of concern that the Yucca Mountain repository may never be licensed, and that interim storage could become permanent.

In Japan, a similar situation ultimately resulted in Japan first shipping its spent fuel to France and the

United Kingdom to be reprocessed and then building a $20 billion domestic reprocessing plant to which spent fuel is now being shipped. In 2006, DoE similarly proposed reprocessing as a "solution" to the U.S. spent fuel problem.

Reprocessing of light-water-reactor fuel is being conducted on a large scale in France and in the United Kingdom. Much of the spent fuel that has been reprocessed has been foreign, notably from Germany and Japan, but since France and the United Kingdom require that the radioactive waste from reprocessing be returned to the country of origin, the need for interim radioactive waste storage in their customer countries was only postponed. In Japan, as part of its agreement to host Japan's domestic reprocessing plant, Amori Prefecture has also agreed to accept for interim storage the reprocessing waste returning from Europe to Japan. Germany and other European countries that were having their spent fuel reprocessed in France, Russia and the UK have decided not to renew their reprocessing contracts and instead plan to store their spent fuel until a geological repository can be sited. France plans to continue reprocessing most of its domestic spent fuel and, like Japan, is storing the resulting radioactive waste at its reprocessing site in La Hague. The United Kingdom is shutting down its reprocessing plants.

The construction of plants to reprocess light-water-reactor spent fuel was originally justified in the 1970s as a way to obtain plutonium to start up liquid-sodium-cooled plutonium-breeder reactors that, in theory, could extract 100 times more energy than current generation reactors from a ton of natural uranium. Breeder reactors were expected to be dominant by the year 2000. The transition to breeder reactors did not occur, however, because their capital costs, and those of reprocessing plants, were much higher than had been projected and because global nuclear generating capacity has grown to only a few percent of the level that was projected in the 1970s. This, along with the discovery of huge deposits of high-grade uranium ore in Australia and Canada, has postponed, for at least a century, concerns about shortages of low-cost uranium. Today, where plutonium is being recycled, it is being recycled as fuel for the light-water reactors (LWRs) from which it was extracted. Even with the cost of the reprocessing ignored as a "sunk cost," plutonium fuel is generally more costly than conventional low-enriched uranium (LEU) fuel.

Worldwide, about half of the plutonium being separated is simply being stockpiled at the reprocessing plants along with the associated high-level waste from reprocessing. In effect, those sites are interim spent-fuel storage sites—except that much of the spent fuel is being stored in separated form. As of 2005, the global stockpile of separated civilian plutonium had grown to 250 tons—sufficient to make more than 30,000 nuclear weapons.

The DoE does not plan to recycle in existing LWRs the plutonium that would, according to its proposal, be separated from U.S. spent fuel. Instead, it proposes that the federal government subsidize the construction of tens of sodium-cooled fast-neutron "burner" reactors—basically, except for changes in their core design, the same sodium-cooled reactors that could not compete economically as plutonium breeder reactors. Plutonium—and, in the future, other less abundant transuranic elements extracted from spent LWR fuel—would be recycled repeatedly through these reactors until, except for process losses, they were fissioned. The principal advantage claimed from doing this would be less long-lived waste per ton of spent fuel, and that the residue from more spent fuel could be stored in the Yucca Mountain repository before a second repository would be required. Such a program would be enormously costly, however. The extra cost to deal with just the spent fuel that has already accumulated in the United States was estimated in 1996 by a U.S. National Academy of Sciences study as "likely to be no less than $50 billion and easily could be over $100 billion." U.S. nuclear utilities have made clear that these extra costs would have to be funded by the federal government. It is quite possible that the program would stop—as previous efforts to commercialize sodium-cooled reactors have—after only one or two "demonstration" reactors have been built. In this case, the reprocessing plant would simply become an interim storage site for the reprocessed spent fuel—as has happened in the United Kingdom and Russia after their breeder-reactor commercialization programs failed.

The French nuclear combine, AREVA, has proposed that it would be less costly to adopt the French approach with a third-generation combined reprocessing and plutonium-fuel fabrication plant in the United States. This would involve recycling the plutonium once in LWRs. The resulting spent "mixed-oxide" (MOX) fuel, which would still contain two-thirds as much plutonium as was used to fabricate it, would then remain indefinitely in interim storage at the reprocessing plant. Thus, once again, the reprocessing plant would serve as a costly type of interim spent-fuel storage.

U.S. Government policy turned against reprocessing after India, in 1974, used the first plutonium recovered by its U.S.-assisted reprocessing program to make a nuclear explosion. Reprocessing makes plutonium accessible to would-be nuclear-weapon makers—

national or subnational—because it eliminates the protection provided by the lethal gamma radiation emitted by the fission products with which the plutonium is mixed in spent fuel.

In early 2006, the DoE originally proposed, as a more "proliferation-resistant" alternative to traditional reprocessing, to keep the reprocessed plutonium mixed with some or all of the minor transuranic elements in the spent fuel. Some of these elements are much more radioactive than the plutonium, but the radiation field that would surround the mix would be one thousand times less intense than the International Atomic Energy Agency (IAEA) considers necessary to provide significant "self protection."

Recently, because of unresolved technical difficulties with fabricating fuel containing some of the minor transuranics, the DoE has sought "expressions of interest" from industry in building a reprocessing plant that would differ from conventional reprocessing only in that it would leave some of the uranium mixed with the plutonium. Pure plutonium could be separated out from this mixture in an unshielded glove box.

In fact, the Bush administration does not argue that any of the variants of reprocessing proposed by the DoE are proliferation resistant enough to be deployed in states of proliferation concern. It has therefore proposed a "Global Nuclear Energy Partnership" in which the weapon states and Japan would provide reprocessing services for other nonweapon states. This proposal has already backfired in stimulating a revival of interest in France in exporting reprocessing technology and in South Korea in acquiring its own national reprocessing capabilities. A similar Bush administration proposal to confine enrichment to states that already have full-scale commercial enrichment plants has similarly stimulated a revival of interest in enrichment in half a dozen nonweapon states.

In comparison, the U.S. policy, which is, in effect, that "we don't reprocess, and you don't need to either," has been much more successful. During the 30-year period it has been in force, no nonweapon state has initiated commercial reprocessing, and seven countries have abandoned their interest in civilian reprocessing. In Belgium, Germany, and Italy domestic developments were more important than U.S. policy. In Argentina, Brazil, South Korea, and Taiwan, however, countries that were interested in developing a nuclear-weapon option, U.S. pressure played a key role. Today, Japan is the only nonweapon state that engages in commercial reprocessing.

The principal alternative to reprocessing, until U.S. spent fuel can be shipped to Yucca Mountain or some other centralized storage, is simply to keep older spent fuel in dry storage on the reactor sites. There is ample space inside the security fence at all U.S. power-reactor sites to store all the spent fuel that will be discharged, even if the reactor licenses are extended to allow them to operate until they are 60 years old. At an operating reactor site, the incremental safety and security risk from dry stored fuel is negligible relative to the danger from the fuel in the reactor core and the recently discharged fuel in the spent fuel pool.

II. INTRODUCTION

In 2006, in response to congressional pressure to start moving spent fuel off U.S. power-reactor sites, DoE proposed U.S. Government-funded reprocessing of the fuel and recycling of the recovered plutonium and minor transuranic elements. If carried through, this proposal would reverse a nonproliferation policy established by the Ford and Carter administrations after India, in 1974, used the first plutonium it extracted as part of a U.S.-supported reprocessing program, to make a nuclear explosion. U.S. policy became to oppose reprocessing where it was not already established and not to reprocess domestically.[1] Four years later, in 1981, the Reagan administration reversed the ban on domestic reprocessing.[2] By that time, however, U.S. utilities had learned that reprocessing would be very costly and were unwilling to pay for it.[3]

The Nuclear Waste Policy Act of 1982 therefore established that, in exchange for revenue from a tax of 0.1 cent per nuclear-generated kilowatt-hour of electricity, starting in 1998, DoE would take spent power reactor fuel from U.S. nuclear utilities and place it in an underground federal geological repository.[4] In 1987, Congress decided to site the first such repository under Yucca Mountain, Nevada.[5] Project management problems and determined opposition by the State of Nevada, however, have delayed the licensing process. Currently, DoE expects to receive a license for the Yucca Mountain repository in 2017 at the earliest.[6] U.S. utilities therefore have been suing DoE for the costs of building on-site dry-cask storage for the spent fuel that would have been shipped to Yucca Mountain on the originally contracted schedule. DoE has informed Congress that the cost of settling these lawsuits is likely to climb to $0.5 billion per year of delay in licensing the Yucca Mountain repository.[7] DoE has refused to share the basis for this estimate because of the lawsuits. The incremental cost for additional storage capacity, after the nuclear power plants have paid for the infrastructure for dry-cask storage (most have already) probably will be somewhat less.[8] In any case, the costs would be about the same if DoE had to pay for off-site storage.

Even if the Yucca Mountain repository had been licensed on time, however, DoE would have faced another problem. When Congress selected Yucca Mountain to be the site of the first U.S. geological spent-fuel repository, it limited the quantity of commercial spent fuel that could be stored there to 63,000 tons until a second repository is in operation.[9] U.S. nuclear power plants will have discharged about 63,000 tons of spent fuel by the end of 2008. DoE is therefore faced with the challenge of siting a second repository at a time when it has not yet succeeded in licensing the first one. The Bush administration has submitted legislation that would remove the 63,000-ton legislated limit. It is believed that the physical capacity of Yucca Mountain is great enough to hold the lifetime output of the current generation of U.S. power reactors and perhaps several times that amount (see below).

Because of the delay in the availability of the Yucca Mountain repository, in 2005 Congress asked DoE to develop a plan for centralized interim storage and reprocessing of U.S. spent fuel. In May 2006, DoE responded with a plan for a "Global Nuclear Energy Partnership" (GNEP) as a part of which DoE would build reprocessing plants and subsidize the construction of tens of fast-neutron reactors to fission the recovered plutonium and other transuranic elements. DoE argues that, if the transuranics are fissioned and the 30-year half-life fission products that generate most of the heat in the resulting waste are stored on the surface for some hundreds of years, then residues from much more spent fuel could be stored in Yucca Mountain.

DoE's Argonne National Laboratory, which provides technical support for DoE's research and development (R&D) program on advanced reprocessing technologies, envisioned GNEP as limited for many years to an R&D program, because the technology for recycling the minor transuranics, americium and curium, is not in hand. Paul Lisowski, DoE's Deputy Program Manager for GNEP, has described transuranic recycle as a "major technical risk area for GNEP."[10] Under congressional pressure to move more quickly, however, DoE issued a request to industry for "Expressions of Interest" in constructing a conventional reprocessing plant and a demonstration fast-neutron reactor as soon as possible. The most likely contractor for construction of the reprocessing plant, the French nuclear conglomerate AREVA, advises the United States to defer recycling anything other than plutonium and to build a larger-capacity version of France's reprocessing and plutonium recycle infrastructure. Specifically, it proposes that the plutonium in recently discharged U.S. spent fuel be recycled once in LWRs and then the resulting spent MOX fuel be stored at the reprocessing plant until the advent of fast-neutron "burner" reactors.[11]

The U.S. House of Representatives insisted, however, that a "first test of any site's willingness to host such a facility is its willingness to receive into interim storage spent fuel in dry casks . . . Resolution of the spent fuel problem cannot wait for the many years required for . . . GNEP [which] will not be ready to begin large-scale recycling of commercial spent fuel until the end of the next decade, and the Yucca Mountain repository will not open until roughly the same time. Such delays are acceptable only if accompanied by interim storage beginning *this decade*" [emphasis added].[12]

Thus the revived interest in the United States in reprocessing is very much entangled in the perceived urgency of starting to move spent fuel off of reactor sites.

The report that follows describes the history of interest in civilian reprocessing, past experience with reprocessing costs, estimates of its likely costs in the United States with and without transmutation of the recovered transuranic elements, and the debate over the relative "proliferation resistance" of alternative fuel cycles. It concludes that a much less costly and proliferation resistant alternative to reprocessing and transuranic recycle would be continued on-site storage of U.S. spent fuel until either Yucca Mountain or some other off-site location is available.

III. HISTORICAL BACKGROUND

Fuel reprocessing was invented during World War II as a way to recover plutonium for nuclear weapons from irradiated reactor fuel. From the 1950s through the 1970s, however, it was expected to play an essential role in civilian nuclear power as well.

The Original Rationale for Reprocessing.

This expectation was based on the belief that deposits of high-grade uranium ore were too scarce to support nuclear power on a large scale based on a "once-through" fuel cycle. The once-through fuel cycle, as realized with the dominant LWR today, involves the production of LEU containing about 4 percent U-235, which is then irradiated until most of the U-235 and about 2 percent of the U-238 have been fissioned, and then is stored indefinitely (see Figure 1.)

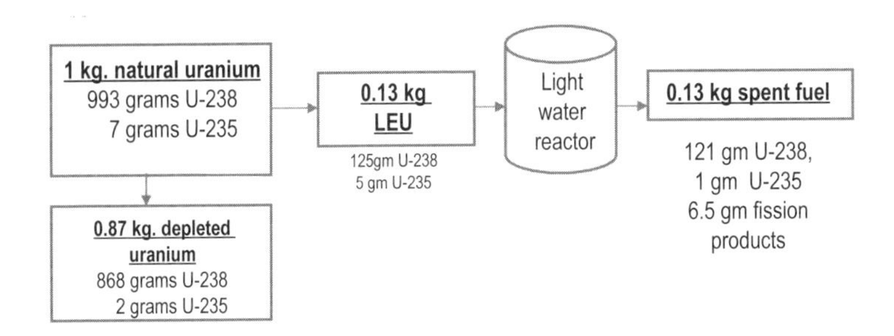

The once-through fuel cycle fissions less than 1 percent of the atoms in natural uranium, but it is less costly and more proliferation resistant than fuel cycles involving reprocessing. If, in the future, reprocessing becomes economical and otherwise acceptable, the uranium that is not fissioned in the once-through fuel cycle will still be available in the depleted uranium and spent fuel.[13]

Figure 1. The Once-Through Fuel Cycle.

This fuel cycle uses most of the fission energy stored in the rare chain-reacting uranium isotope, U-235, which makes up 0.7 percent of natural uranium. Atom for atom, however, the U-238 atoms, which make up virtually all of the remaining 99.3 percent of natural uranium, contain as much potential fission energy. If it were possible to fission the U-238, the amount of energy releasable from a kilogram of natural uranium therefore would be increased about 100-fold.

Plutonium breeder reactors. A month after the first reactor went critical under the stands of the University of Chicago's football stadium, Leo Szilard, who first conceived of the possibility of a nuclear chain reaction, invented a reactor that could efficiently tap the energy in U-238 by turning it into chain-reacting plutonium. In a sodium-cooled reactor, a chain reaction in plutonium would be sustained by "fast" neutrons that had not been slowed down as much by collisions with the sodium coolant as neutrons are in collisions with the light hydrogen atoms in the cooling water of conventional reactors. Plutonium fissions by fast neutrons produce enough neutrons so that it is possible on average to convert more than one U-238 atom into plutonium per plutonium atom destroyed.[14] Such reactors are called plutonium "breeder" reactors. Alternatively, they can be thought of as U-238 burner reactors.

Being able to exploit the energy stored in the nucleus of U-238 would make it possible to mine ores containing about 1 percent as much natural uranium as could be economically mined for the energy in U-235 alone. Indeed, even the 3 grams of uranium in a ton of average crustal rock, if fissioned completely, would release almost 10 times as much energy as is contained in a ton of coal.[15] The nuclear-energy pioneers therefore talked of breeder reactors making it possible to "burn the rocks" and thereby create a source of fission energy that could power humanity for a million years.

The growth of global nuclear-power capacity slowed dramatically in the 1980s, however, (see Figure 2) and huge deposits of rich uranium ore were discovered in Australia, Canada, and elsewhere. As a result, the long-term trend of natural-uranium costs has been down rather than up (see Figure 3). Concerns about uranium shortages linger on today in arguments that nuclear power based on a "once-through," LEU fuel cycle is not "sustainable." But such concerns about the inadequacy of the world's uranium resources have shifted to far beyond 2050.[16] In any case, depleted uranium and spent fuel can be stored so as to be available in the event that it becomes cost-effective to "mine" them for the energy in their uranium-238.

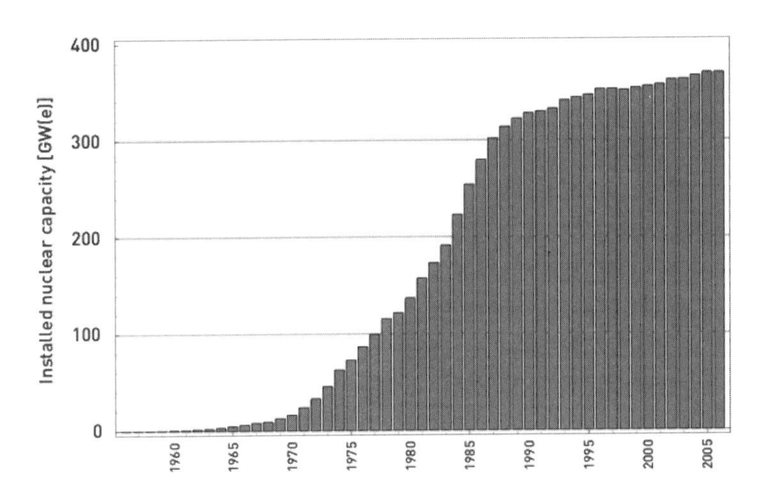

Global nuclear generating capacity grew rapidly in the 1970s, leading to concerns that the supply of natural uranium might not be able to keep up with the increasing demand, but growth slowed in the 1980s as a result of the high capital costs of nuclear-power plants, the slowing growth in overall demand for electric power and the Chernobyl nuclear accident of 1986.[17]

Figure 2. Global Nuclear Generating Capacity.

At the same time, the differences between the capital and operating costs of water and sodium-cooled reactors have remained discouragingly large. Many experimental and demonstration breeder reactors have been built around the world but none has been a commercial success.[18]

Because of its compact core, Admiral Hyman G. Rickover, the father of the U.S. nuclear navy, had a sodium-cooled reactor built for the second U.S. nuclear submarine, the *Seawolf*. After sea trials in 1957, however, he had the reactor replaced by a pressurized water reactor. His summary of his experience with the sodium-cooled reactor pretty aptly characterizes the problems that have been subsequently experienced

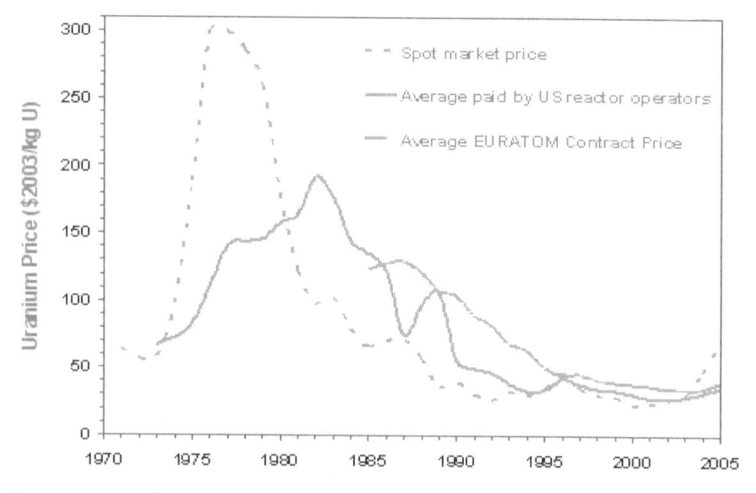

Average and spot uranium prices in constant 2003 dollars, 1971-2005.[19]

Figure 3. Average and Spot Uranium Prices.

in attempts to commercialize sodium-cooled breeder reactors. These reactors are "expensive to build, complex to operate, susceptible to prolonged shutdown as a result of even minor malfunctions, and difficult and time-consuming to repair."[20]

In anticipation of a need for large quantities of separated plutonium to provide startup cores for the breeder reactors, however, commercial reprocessing of spent LWR fuel was launched in the 1960s. Spent LWR fuel contains about 1 percent plutonium. Civilian pilot and full-scale reprocessing plants have been built in eight countries.[21]

Growing stockpiles of separated civilian plutonium. In the absence of significant breeder-reactor capacity, some countries—notably France and Germany—have been recycling their separated plutonium back into LWR fuel. The cost of fabricating MOX plutonium-uranium fuel for LWRs has been greater, however, than the value of the LEU fuel that has been saved.[22] As a result, there is no commercial demand for plutonium as a fuel and large stockpiles have accumulated at the reprocessing plants, along with the fission-product waste from which the plutonium was separated. The United Kingdom and Russia have stockpiled all the plutonium that they have been separating from their own spent fuel (and, in Russia's case, also from the spent fuel that Eastern and Central European utilities have been shipping to Russia for reprocessing). Japan's separated plutonium has accumulated at the French and U.K. reprocessing plants because local government opposition in Japan has delayed its plutonium recycle program for a decade.[23]

Based on declarations of civilian plutonium stocks to the IAEA, the global stock of separated civilian plutonium has been growing by an average of 10 tons per year since 1996 and was about 250 metric tons as of the end of 2005 (see Table 1). This stockpile is approximately the same size as the global stockpile of plutonium that was produced for weapons during the Cold War. About 100 tons of Russian, U.S., and U.K. weapon plutonium have been declared excess, increasing the global stockpile of excess separated plutonium still further.

As an energy resource, the world stockpile of separated civilian plutonium is not huge. It could fuel the world's fleet of power reactors for less than a year. In terms of weapon equivalents, however, it *is* huge. Using the IAEA's 8-kg weapon equivalent, the 350 tons of civilian and excess weapons plutonium could be converted into 40,000 first-generation (Nagasaki-type) nuclear weapons. In 1998, a Royal Society report observed that the possibility that the United Kingdom's very large stockpile of separated civilian plutonium "might, at some stage, be accessed for illicit weapons production is of extreme concern."[24] If this is a concern in the United Kingdom, it should be a concern in any country with significant quantities of separated plutonium.

Country	Civilian Stocks (end of 2005)	Military Stocks Declared Excess
Belgium	3.3 (2004) (+0.4 tons in France)	--
China	0	0
France	81 (30 tons foreign owned)	0
Germany	12.5 (+ 15 tons in France & U.K.)	--
India	5.4	0
Japan	5.9 (+38 tons in France & U.K.)	--
Russia	41	34-50
Switzerland	Up to 2 tons in France & U.K.	0
U.K.	105 (27 foreign owned) (+ 0.9 tons abroad)	4.4
U.S.	0	54
TOTALS	≈250 tons	92-108

Table 1. Global Stocks of Separated Civilian and Excess Military Plutonium.[25] (Metric Tons)

Why Reprocessing Persists.

The United Kingdom plans to end its reprocessing by 2012.[26] But France continues, Japan put a big new reprocessing plant into operation in 2006, and the Bush administration has proposed that the United States launch a domestic reprocessing program. Why, in the face of adverse economics, does civilian reprocessing persist?

NIMBY pressures. Reprocessing continued in Western Europe and Japan in the 1980s and 1990s in part because of a combination of local political pressures to do something about the problem of spent fuel accumulating at power-reactor sites and not-in-my-backyard (NIMBY) political opposition elsewhere to geological repositories or central interim storage facilities for spent fuel. Reprocessing provided an interim destination for the spent fuel.

German and Japanese nuclear utilities largely financed the French and British multi-billion-dollar commercial reprocessing facilities.[27] Their respite was only temporary, however, because the reprocessing contracts provided that the solidified fission-product waste would be shipped back to the countries of origin. Germany's anti-nuclear movement finally succeeded in persuading the SPD-Green coalition government to stop reprocessing and eventually phase out nuclear power in Germany and, in exchange, agreed to accept the construction of dry-cask interim spent-fuel storage at the reactor sites until the site of a geological repository could be settled.[28]

Japan's nuclear utilities went down a different route. They persuaded the rural Amori Prefecture to store for 50 years the radioactive waste being returned from Europe as part of an agreement in which the prefecture accepted a large reprocessing plant in return for receiving large payments from a central fund. Japan's nuclear utilities now are shipping their spent fuel to the Rokkasho reprocessing plant. The separated plutonium and high-level waste will be stored there. The high level waste, at least, will stay there until a geological repository can be opened—hopefully within the promised 50 years. The plutonium will be added to Japan's existing 40-ton stockpile of separated plutonium that is eventually to be recycled in MOX fuel.[29]

The Bush administration's reprocessing proposal. U.S. nuclear utilities, too, have been unable to ship their accumulating spent fuel off their reactor sites. As noted above, the reason is delays in the licensing of DoE's proposed geological repository under Yucca Mountain, Nevada. U.S. utilities therefore have been suing DoE for the costs of building additional on-site dry-cask storage.

In 2005, in order to stop these accumulating lawsuits, the U.S. Congress asked DoE to develop a plan for centralized interim storage and reprocessing of U.S. spent fuel.[30] In May 2006, DoE responded with a plan for building reprocessing plants. These reprocessing plants would separate spent LWR fuel into four streams: uranium, plutonium mixed with the other transuranic elements (neptunium, americium, and curium); the 30-year-half-life fission products strontium-90 and cesium-137; and other fission products. This is the so-called UREX+ fuel cycle (see Figure 4).

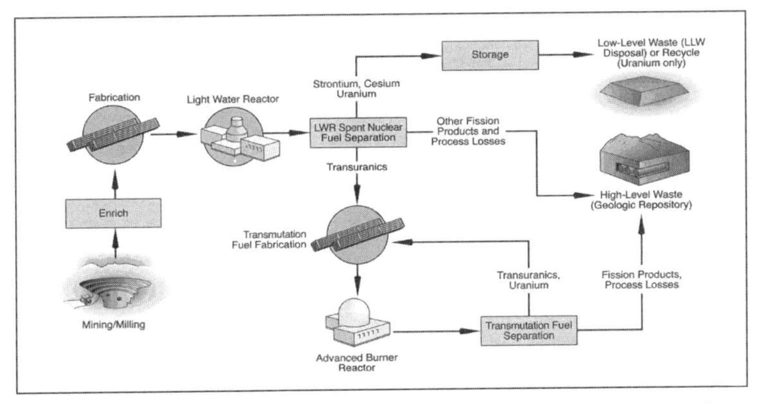

The reprocessing plant (designated here as "LWR Spent Nuclear Fuel Separation") would be built as soon as possible. The reactors shown here as "Advanced Burner Reactors" would be fast-neutron reactors. Only one would be built at the same time as the reprocessing plant. Others would be built on an unspecified time schedule. After reprocessing, the 30-year half-life isotopes, cesium-137 and strontium-90, which dominate the radiological hazard until they decay away, would be placed in interim surface storage for some hundreds of years. This raises the question as to why the unreprocessed spent fuel should not be remain in interim storage until fast-neutron reactors actually are built in significant numbers.[31]

Figure 4. The Department of Energy's May 2006 Proposal for Reprocessing U.S. Spent Fuel and Fissioning the Transuranics.

The transuranic elements would be recycled in a hypothetical future generation of fast-neutron "burner" reactors until—except for losses to various waste streams—the transuranics were fissioned. The designs of the burner reactors would be adapted from the sodium-cooled reactors that previously were to be commercialized as plutonium-breeder reactors, only with the plutonium breeding uranium blankets around their cores removed. The uranium would be stored or disposed of as waste. The strontium-90 and cesium-137 would be placed into interim surface storage for some hundreds of years—presumably at the reprocessing plant. Only the residual wastes after the separation of these three streams would be placed in the Yucca Mountain repository.

By removing in each cycle 99 percent of the strontium-90 and cesium-137 and of the transuranic elements, the main sources of radioactive decay heat in the spent fuel on century and millennial scales, respectively, the long-term temperature increase of the rock around the disposal tunnels under Yucca Mountain per ton of spent fuel would be decreased about 20-fold. The residue from 20 times as much spent fuel therefore could be emplaced in the mountain before a new repository would have to be sited.[32] The political resistance to the siting of the Yucca Mountain repository has been so fierce that this is considered by DoE to be a major long-term advantage of the proposed UREX + fuel cycle and a prerequisite for nuclear power to have a long-term future in the United States.

The current limit on the capacity of Yucca Mountain, however, is not physical but *legislated*. When Congress selected Yucca Mountain as the nation's first geological radioactive waste repository, it wished to reassure Nevada that it would not have to carry this burden alone. As already noted, it therefore limited the quantity of commercial spent fuel or reprocessing waste that can be stored there to 63,000 tons "until such a time as a second repository is in operation." This amount of spent fuel will have been discharged by U.S. reactors by 2008. Hence the dire warnings of the necessity to site repositories in additional states. In order to deal with this problem, the Bush administration has proposed to lift the legislated limit on the amount of spent fuel that can be stored in Yucca Mountain.[33]

The federal government has not come to its own conclusion about what the *physical* capacity of Yucca Mountain might be. Using federal studies made as part of the licensing process for the repository, however, the utility industry's Electrical Power Research Institute estimates that there is enough capacity in the surveyed areas of Yucca Mountain to store 260,000 -570,000 tons of spent fuel—and perhaps more. This is two to five times as much as the current generation of U.S. power reactors are expected to discharge over their lifetimes.[34]

Because of the delay in licensing the repository and the utility lawsuits, however, the Congressional Appropriations Subcommittees that fund DoE have been pressing DoE to begin moving spent fuel off power reactor sites. In part at least in response to this pressure, on August 7, 2006, DoE announced that it was considering building a 2000-3000 ton per year spent-fuel reprocessing plant based on the existing technology being used in France, and a 2000 MWt (thermal) sodium-cooled fast-neutron reactor of the pool-type design used for France's failed Superphénix reactor. The reprocessing plant would be modified so that some of the uranium in the spent fuel would remain mixed with the plutonium. In this way, DoE would honor its commitment to make reprocessing more "proliferation resistant." Plutonium can be separated out of such a mixture very much more easily, however, than from spent fuel (see Section V). The fast reactor would be fueled initially by "conventional fast reactor fuel," i.e., a mix of plutonium and uranium.[35] In January 2007, DoE announced that it planned to lay the basis for a decision by the Secretary of Energy to launch this program "no later than June 2008," i.e., before President Bush leaves office.[36]

Reprocessing 2000-3000 tons of LWR spent fuel would separate 24-36 tons of plutonium per year.[37] By comparison, France's failed 3000 MWt Supérphenix, even operating on a once-through fuel cycle, would have annually irradiated only about 2 tons of plutonium.[38] In effect, unless DoE adopts the French strategy of recycling MOX in LWRs, its reprocessing initiative would, for the foreseeable future, transform almost all spent fuel shipped from U.S. nuclear-power-reactor sites into separated plutonium and high-level waste stored at a reprocessing site. The compelling reason for DoE initiative, therefore, appears to be, as in Japan, to provide an alternative destination for spent fuel until a geological radioactive waste repository becomes available.

DoE's reprocessing proposals are controversial both because of their cost and their impact on U.S. nonproliferation policy. We discuss these issues in the next two sections.

IV. REPROCESSING AND RECYCLE COSTS

We consider the costs for two scenarios:

1. DoE's May 2006 scenario in which all of the transuranics in U.S. light-water spent fuel would be separated and fissioned in fast-neutron reactors in order to increase the number of reactor-years of radioactive waste that can be accommodated in Yucca Mountain. Although DoE has never mentioned it in connection with its current proposal, these costs were examined in depth in a massive National Academy of Sciences study that was commissioned by DoE in the early 1990s and published in 1996.[39]

2. The cost and benefits of doing what is done in France, which is to reprocess spent LWR fuel, mix the

separated plutonium with depleted uranium to make MOX fuel for LWRs, and then store the spent MOX fuel. AREVA, the French nuclear conglomerate, has launched a major effort to convince DoE to follow this route, including by funding a study that claims that reprocessing would not be much more costly in the United States than building a second geological repository for spent fuel.[40]

The 1996 Study by the U.S. National Academy of Sciences.

The 1996 U.S. National Academy of Sciences study estimated the extra cost of a separations and transmutation program for the first 62,000 tons of U.S. spent fuel, relative to the cost of simply storing the spent fuel in a repository, as "likely to be no less than $50 billion and easily could be over $100 billion" (1996$).[41] For the estimated lifetime discharges of the current generation of U.S. LWRs (101,000 to 129,000 tons, see Figure 5), this cost would be approximately double.

Currently, U.S. nuclear utilities are paying into DoE's Nuclear Waste Fund 0.1 cents per kilowatt-hour in exchange for DoE taking responsibility for disposing of their spent fuel. Assuming that the average amount of fission energy released in the first 62,000 tons of U.S. spent fuel was 40,000 megawatt-days per ton and taking the heat-to-electric energy conversion efficiency of an average nuclear power plant to be one-third, this would translate into about $20 billion. Even including interest, this fund would not be able to cover both the estimated $50 billion cost of the Yucca Mountain repository and a $100 billion separations and transmutation program.[42] Spokesmen for the nuclear utilities have made clear that they will not pay for the extra costs of a reprocessing plant or fast-neutron reactors.[43]

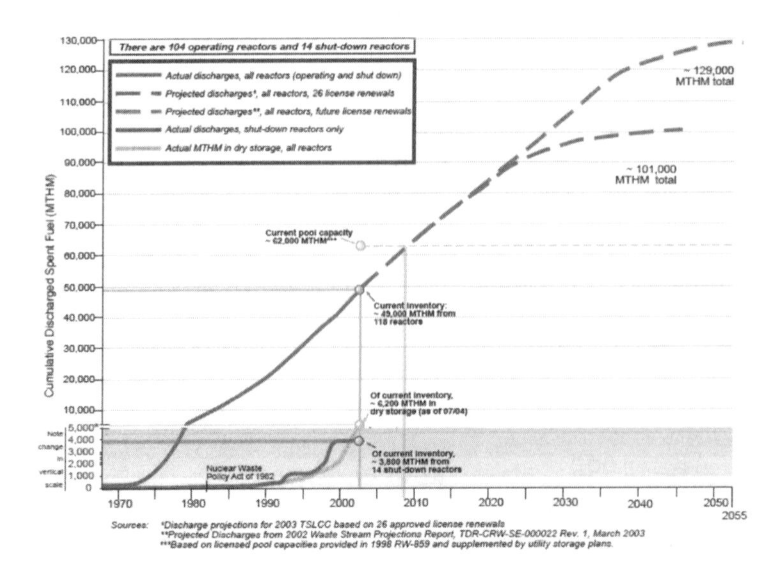

Projections of the total amount of spent fuel to be discharged by the current generation of U.S. power reactors depend upon what fraction of the reactors have their licenses extended to 60 years.[44]

Figure 5. Projections of the Total Amount of Spent Fuel To Be Discharged by the Current Generation of U.S. Power Reactors.

It is conceivable that the U.S. Congress might fund the launch (although perhaps not the completion) of a federally funded reprocessing plant costing tens of billions of dollars, but it seems unlikely that it would provide a subsidy of on the order of a billion dollars each for the construction of 40-75 fast-neutron reactors to fission the transuranics being produced by 100 gigawatts electrical (Gwe) of LEU-fueled LWRs.[45]

The great cost of DoE's proposed program and the fact that it proposes to store the most dangerous isotopes in the spent fuel[46] on the surface for hundreds of years may eventually increase the appeal of interim storage without reprocessing.

The AREVA Study of the Cost of Recycling Separated Plutonium in MOX.

In July 2006, the Boston Consulting Group published a report, *Economic Assessment of Used Nuclear Fuel Management in the United States*. The report was commissioned by the French nuclear combine, Areva, and is based on proprietary data and analysis provided by Areva. The report will therefore be referred to below as the "Areva study."

The report proposes that AREVA build for the U.S. Government both a spent-fuel reprocessing plant with a 2,500 ton-per-year capacity and a mixed-oxide fuel fabrication plant to recycle the separated plutonium back into LWR fuel. It argues that the cost would approximately equal the savings from the United States being able to delay a second repository by 50 years.[47] Given the similarities of this proposal to DoE's request 2 weeks later for expressions of interest in building a reprocessing plant with a capacity of 2,000-3,000 tons a year, it is worth examining the Areva report's analysis. Below, we examine the basis of its central conclusions that:

1. AREVA could build and operate a reprocessing plant and MOX fuel fabrication plant much more cheaply for the U.S. Government than it did in France; and,

2. French-style reprocessing and plutonium recycle would postpone the need of a second U.S. repository.

Finally, we will summarize the results of a French Government analysis of the net costs of plutonium recycle in France.

Lower Costs in the United States than in France? The

AREVA study asserted that reprocessing and MOX fuel-fabrication plants could be built in the United States more cheaply than the corresponding smaller-capacity facilities it built in France.[48] The capital cost of the French complex was revealed to be about $18 billion in 2006 dollars, not including interest charges during construction. The study also asserted that the plants could be operated for about $0.9 billion per year—about one-third the operating cost shown for the smaller complex in France.

France's spent-fuel reprocessing complex on Cap de La Hague in northern France. Its plutonium fuel fabrication facility is in southern France, requiring regular long-distance truck shipments of separated plutonium.[49]

Figure 6. France's Spent-Fuel Reprocessing Complex on Cap de La Hague.

Thus far, however, DoE-AREVA combination has resulted in much *higher* costs in the United States than in France. DoE has contracted with AREVA to build a MOX fuel fabrication plant to deal with 34 tons of excess U.S. weapon plutonium at a rate of 3.5 tons per year.[50] Measured in terms of MOX fuel tonnage, this is about one-fifth the capacity of the plant that would be required to take the plutonium output of 2,500 ton/year reprocessing plant.[51] The original estimated cost of DoE's MOX-fuel facility presented to Congress in 2002 was $1 billion. By July 2005, 3 years later, the estimated cost had ballooned to $3.5 billion, and the project was 2.5 years behind schedule.[52] Such cost overruns and delays are typical for DoE projects.[53]

Would French-style reprocessing postpone the need for a second repository? For the non-reprocessing alternative to its proposal, the AREVA study assumed that the physical capacity of Yucca Mountain is 120,000 tons of spent LEU fuel. As indicated above, the capacity is likely to be much larger. Using AREVA's assumption, however, at the current rate of discharge of spent fuel by U.S. power reactors, (about 2,000 metric tons of heavy-metal content per year) the Yucca Mountain repository would be fully subscribed by 2040. Fuel discharged later could not be loaded into a repository until it had cooled for 25 years, i.e., till 2065, but the AREVA study assumed that, already in the year 2030, the United States would have to start spending $0.4 billion a year on a $45-50 billion second repository.[54]

Americium-241 (Am-241), which forms from the decay of 14-year half-life plutonium-241, dominates the heat output of LEU spent fuel during the period from 100 years to 2,000 years after discharge. In AREVA's proposal, the Am-241 would go into the high-level reprocessing waste and be emplaced in Yucca Mountain.

To minimize the buildup of Am-241 in the spent fuel and thereby the amount of Am-241 in the high level waste, the AREVA study assumes that, after the reprocessing plant is completed, spent fuel would be reprocessed within 3 years. This would reduce the heat load from the associated high-level waste to the point where the waste from 230,000 tons of spent fuel could be stored in Yucca Mountain plus 50,000 tons of unreprocessed pre-2003 spent fuel—more than doubling the amount of spent fuel that could be dealt with before a second repository would have to be established.[55]

The AREVA study is able to postpone the need of a second U.S. repository beyond the study's time horizon, however, only because it assumes that the spent MOX fuel would remain indefinitely in interim storage at the reprocessing plant. There would be no delay in the need for a second repository had it been assumed that the spent MOX fuel, too, would be emplaced in Yucca Mountain. Although reprocessing and plutonium recycle consolidates the plutonium from roughly eight tons of spent LEU fuel into one ton of fresh MOX fuel, the total amount of plutonium in the spent MOX fuel is still two-thirds as great as in the original eight tons of LEU spent fuel. Furthermore, because of a shift toward a hotter mix of plutonium and other transuranics, the amount of heat that the ton of MOX spent fuel would deliver into the mountain during the first crucial 2,000 years would be almost exactly the same as would have been delivered by the eight tons of spent LEU fuel. This is why the AREVA study states that "[D]isposal of MOX [in a geological repository] is not considered to be a viable option."[56] Indeed, the French Government has concluded that spent MOX fuel would have to be stored from 150 years to "centuries" before it cooled enough to be emplaced in a geological repository.[57]

A complete cost analysis would have dealt with cost of an alternative way of disposing of the spent

MOX fuel. DoE proposes that the plutonium should be recycled repeatedly in fast-neutron reactors until it is completely fissioned. If this were done after one recycle in LWRs had reduced the amount of plutonium by one-third, only 23-44 GWe of fast reactor capacity would be required to fission the plutonium left in the once-recycled LWR MOX fuel.[58] This is down from the 40-75 GWe calculated above for DoE's scenario, in which the plutonium is fed directly into sodium-cooled burner reactors. But the cost would still be huge. The AREVA report assumes that sodium-cooled reactors would cost 20 percent more per unit of generating capacity than LWRs.[59] The only *full-sized* sodium-cooled ever built, France's *Superphénix,* cost about three times as much as a LWR of the same capacity.[60] In series production, the cost could come down. LWRs are estimated to cost $2 billion per GWe. The extra capital cost for buying sodium-cooled reactors therefore would be $9-18 billion if AREVA's 20 percent estimate were true and $46-90 billion if the cost of a breeder were twice that of an LWR. Tens of billions more would be required for the infrastructure to fabricate and reprocess the sodium-cooled reactor fuel.

The French Government's estimate of the cost of reprocessing in France. The AREVA study did not reveal the cost of reprocessing and plutonium recycle in France, but these costs were published in a study done by the French Government in 2000. This study also estimated the costs of alternative fuel cycles for France's current fleet of power reactors.

Shown in the Appendix are the results for four scenarios: three treated in the French Government report and one extrapolated from the results of those calculations:

1. One hundred percent of the LEU spent fuel discharged from France's LWRs in a 45-year average

operational lifetime would be reprocessed (the extrapolated scenario). The separated plutonium would be recycled in MOX fuel *once*—i.e., spent MOX fuel would not be reprocessed within the time frame of the study.

2. About two-thirds of the LEU fuel would be reprocessed, and the plutonium recycled once (the current plan).

3. Reprocessing would end in 2010. This would amount to reprocessing 27 percent of the spent LEU fuel expected to be discharged in the reactors' lifetimes.

4. A retrospective scenario in which France was assumed not to have built its reprocessing and plutonium recycle infrastructure but instead would have deposited its spent fuel directly in an underground repository as is current U.S. policy.

The cost estimates are summarized in Table 2. It will be seen by comparing the 100-percent-reprocessing with the no-reprocessing scenarios that reprocessing all of the LEU fuel would double the cost of the back end of France's fuel cycle. The net increase is 80 percent when the savings in natural uranium and enrichment associated with the use of the MOX fuel are taken into account.

	Percentage of Spent LEU Fuel Reprocessed			
	100% (Derived scenario)	67%	27% (Reprocessing ends in 2010)	No Reprocessing
Back end costs	84	74	61	41
Front end cost savings from plutonium recycle	-10	-8	-2	0
Net costs	74	66	59	41

Table 2. Spent-Fuel Disposal Costs in Four Scenarios for the French Fuel Cycle.[61] (Billions of 2006 $, 58,000 tons of spent fuel)

V. ATTEMPTS TO MITIGATE THE IMPACT ON U.S. NONPROLIFERATION POLICY

Following India's 1974 nuclear explosion, which used civilian plutonium separated with U.S.-provided technology, the United States reversed its policy of encouraging reprocessing and plutonium recycle worldwide. U.S. policy became, in effect, "We don't reprocess, and you don't need to either." Since 1977, when Japan put its Tokai-mura pilot plant into operation, no nonweapon state has begun civilian reprocessing. During that same period, Argentina, Belgium, Brazil, Germany, and Italy shut down their pilot reprocessing plants, and South Korea and Taiwan abandoned their laboratory-scale reprocessing research. Japan remains the only nonweapon state that reprocesses. In Europe, countries have abandoned reprocessing primarily as a result of anti-nuclear movements and the high cost of reprocessing. Outside Europe and Japan, however, U.S. anti-reprocessing policy has played a key role in stopping programs that were covers for countries that were interested in following India's example and using a civilian reprocessing program as a cover for developing a nuclear-weapon option.

The Bush administration has responded in two ways to concerns that a new U.S. reprocessing initiative would undermine this very successful nonproliferation policy:

1. DoE is developing reprocessing technologies that do not separate out pure plutonium.

2. The Bush administration has proposed that reprocessing and uranium enrichment be confined to "countries that already have substantial, well-established fuel cycles."[62]

"Proliferation Resistant" Fuel Cycles—The Saga of UREX+.

The reprocessing technology currently used worldwide has the acronym PUREX for Plutonium and URanium EXtraction. It was originally developed by the United States to extract pure plutonium for the U.S. nuclear-weapons program.[63] It is therefore difficult to claim that this technology is proliferation resistant, and DoE has not done so.

In fact, the revival of U.S. interest in reprocessing was launched by the 2001 report of Vice President Cheney's National Energy Policy Development Group, which recommended that "the United States should reexamine its policies to allow for research, development and deployment of fuel conditioning methods (such as pyroprocessing) that reduce waste streams and enhance proliferation resistance."[64]

Pyroprocessing is a reprocessing technology developed by Argonne National Laboratory (ANL) for recycling the metal fuel used in its Experimental Breeder Reactor II.

Another reprocessing technology would be required, however, to separate transuranics from the uranium-oxide fuel used in LWRs. For this purpose, ANL proposed what it called UREX+, named to denote the fact that pure uranium is extracted. The transuranics are extracted in various combinations in different variants of UREX+. In fact, a series of versions of UREX+ have been proposed.

Plutonium Plus Neptunium.

The first version of UREX+ proposed by Argonne (UREX+2)[65] would keep the plutonium mixed with neptunium.[66] There is, however, typically only about 8 percent as much neptunium as plutonium in spent fuel. Furthermore, neptunium is less radioactive than plutonium and is as good a weapons material as the U-235 used in the Hiroshima bomb. At best, the effect of leaving the neptunium mixed with the plutonium would be to dilute the plutonium slightly. The mix could be used directly to make weapons, or the plutonium could be extracted in the same type of glove box that would be used to handle pure plutonium.

Unseparated transuranics (*UREX+1a*). The second iteration of UREX is the GNEP fuel cycle proposed by DoE in May 2006. It would leave all the transuranics unseparated. Plutonium would still constitute more than 80 percent of the mix. The mix would be about 100 times more radioactive than pure plutonium but would still produce only about 0.1 percent of the intensity of penetrating radiation that would be required to make it "self-protecting" by the IAEA's standard (see Figure 7).[67] Enough plutonium for a few bombs could be separated in a glove box without the workers receiving a large radiation dose. For an industrial-scale operation in which workers were exposed to this material year around, however, shielding and remote handling would be required to keep down occupational radiation doses. This is why "addition of minor [transuranics] or fission products to recycled plutonium will increase significantly the costs of fuel fabrication and transportation."[68]

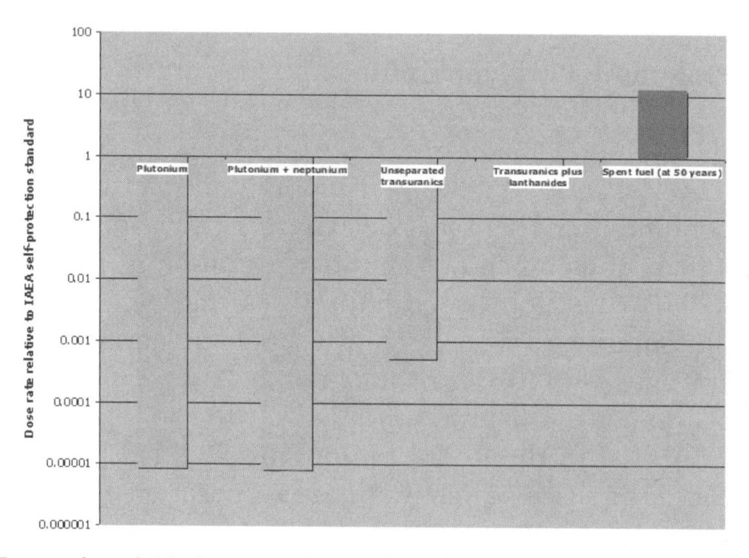

Factors by which dose rates from 1-kg spheres of transuranic metal produced by various versions of UREX+ fall short of the IAEA threshold for self protection (1 Sievert or 100 rems per hour at one meter). For example, the dose rate from unseparated transuranics is about 0.001 of the self-protection standard.[69]

Figure 7. Factors by which Dose Rates from 1-kg Spheres of Transuranic Metal Produced by Various Versions of UREX+ Fall Short of the IAEA Threshold for Self-Protection.

Unseparated transuranics mixed with lanthanide fission products (UREX+1). Argonne responded to criticisms of the lack of proliferation resistance of UREX+1a by proposing yet another variant in which one class of fission products, the lanthanides, would remain mixed with the transuranics until the mix was transported to a sodium-cooled "burner reactor" site (see Figure 8). Although still not meeting the IAEA's self-protection standard, the gamma-radiation level from the mix would be higher than for the other UREX+ fuel cycles considered earlier. It would be highest for material separated from recently discharged spent fuel, since the longest-lived significant lanthanide, Europium-154, has a half-life of only 8.8 years. At the burner-reactor sites, the lanthanides would be stripped out in a final stage of reprocessing, and the transuranic fuel would be fabricated. Thus each burner reactor site would have its own final-stage reprocessing and fuel-fabrication plant. This would compound the problem of the high cost of the separations and transmutation approach. Indeed, the complexity of this proposal approaches that of a Rube Goldberg cartoon.[70]

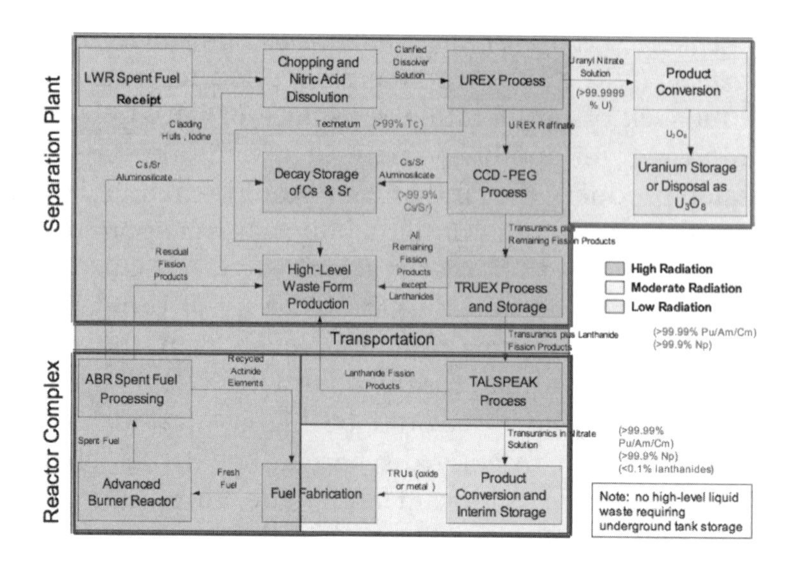

The top box describes the various stages of the reprocessing plant and includes provisions for surface storage for hundreds of years of the two most hazardous fission products, cesium-137 and strontium-90, both of which have half-lives of about 30 years. The box at the bottom describes one of many proposed "burner-reactor" complexes. Each reactor site would have a facility to carry out the final stage of the UREX+ reprocessing (TALSPEAK). It would also have a fuel-fabrication facility and a spent-fuel reprocessing facility for the burner reactors. The enormous number of fuel processing facilities in this proposal would make it much more costly even than the separations and transmutation arrangements analyzed in the 1996 National Academies study.[71]

Figure 8. The Version of UREX+ Proposed by Argonne National Laboratory in March 2006.

Safeguards problems.[72] IAEA has been unable to reduce statistical measurement uncertainties below about 1 percent for traditional PUREX reprocessing, which produces pure plutonium. To prevent frequent false alarms, a 1 percent measurement uncertainty requires raising the alarm threshold to about 3 percent.[73]

Three percent of the 24 tons of plutonium discharged annually by U.S. power reactors would amount to 760 kilograms, enough for about 100 Nagasaki bombs.

Unfortunately, the Argonne proposals to make reprocessing more "proliferation resistant" by adding radioactive materials to the plutonium also would make it more difficult for both national and international monitors to detect plutonium diversion.

Plutonium is ordinarily detected and measured by the penetrating radiation that it emits. It fissions spontaneously at a low rate, emitting neutrons (about half a million per kilogram per second for reactor-grade plutonium). The neutrons can be detected through substantial shielding. Leaving plutonium mixed with other transuranics makes neutron measurements much less useful, however. The Curium-244 in spent fuel, in particular, emits 100 times as many neutrons.[74] As a result, an uncertainty of only *1 percent* in the Curium-244 would mask the loss in neutron signal due to the removal of *all* the plutonium.

All the plutonium isotopes also emit characteristic gamma rays. These gamma rays are much less penetrating than the neutrons, however. Large corrections must therefore be made for shielding and self-shielding of the fissile material. For this reason, gamma measurements are almost useless for quantitative assays of bulk inhomogeneous mixtures.

Back to MOX. Most recently, after learning that UREX+ was still very much in the conceptual stage and that techniques for fabricating fuel containing americium and curium had not yet been developed, DoE decided to explore the possibility of starting with a slight modification of a PUREX plant. In its August 2006 "request for expressions of interest," it specified only that the reprocessing plant "products are not pure plutonium."[75] This was only a few weeks after AREVA had proposed COEX, a variant of PUREX in which the plutonium would not be fully separated from the uranium.[76] Of course, once again, the plutonium could be easily separated from the COEX mix in a glove box.

Proposal to Restrict Reprocessing to the Nuclear-Weapon States Plus Japan.

Despite its R&D initiatives to make reprocessing more "proliferation resistant," DoE has never suggested that the improvement could be great enough for reprocessing to be acceptable in states of proliferation concern. Indeed, in its May 2006 presentation of its GNEP proposal, DoE included the Bush administration's February 11, 2004, proposal to deny enrichment and reprocessing technologies "to any state that does not already possess full-scale, functioning enrichment and reprocessing plants," and instead to offer such states reliable access to LEU and reprocessing services.[77]

The idea that other countries can be permanently barred from acquiring enrichment and reprocessing plants has not gained international acceptance, however. An international panel of experts convened by IAEA found that "there is a consistent opposition by many [non-nuclear weapons states] to accept additional restrictions on their development of peaceful nuclear technology without equivalent progress on disarmament."[78]

This issue is currently joined primarily with regard to the assertion by nonweapon states of their rights to have national uranium-enrichment plants. Since the Bush administration's 2004 proposed ban on additional countries acquiring enrichment plants, six nonpossessing countries have expressed increased interest in acquiring them.[79] The U.S. GNEP proposal has, however, already revived interest in reprocessing in South Korea,[80] and AREVA has floated the idea of exporting the plant that it is designing for the American market to a number of nonweapon states that do not currently reprocess.[81]

France, the United Kingdom, and Russia already have been providing reprocessing services to foreign countries, but France and the United Kingdom have lost virtually all of their foreign customers. Russia has kept a few because, unlike France and the United Kingdom, it has been willing to keep the plutonium and radioactive waste it recovers from its foreign customers' spent fuel.

In effect, Russia has been providing permanent storage for foreign spent fuel—although with the fuel separated into three components: uranium, plutonium, and high-level waste. Under these conditions, its customers have been happy for Russia to take their spent fuel, whether it reprocesses it or not. Indeed, while Russia has been reprocessing the spent fuel from first-generation East European VVER-440 reactors at its Mayak facility in the Urals, it has been storing the spent fuel from second- generation Soviet-designed VVER-1000 reactors in a second closed nuclear city, Zheleznogorsk, Siberia.

VI. THE ALTERNATIVE: DRY-CASK SPENT FUEL STORAGE

In the Sections II and III of this chapter, we discussed how pressure from U.S. nuclear utilities on DoE to remove spent fuel from their reactor sites and the unwillingness of U.S. state governments to host off-site interim storage have stimulated DoE interest in federally-funded reprocessing and recycle of transuranics. In Sections IV and V, we discussed the huge costs of such a program and the weaknesses of proposals to make reprocessing "proliferation resistant." In this section, we discuss whether, considering the alternatives, interim storage of unreprocessed spent fuel on the power-reactor sites may after all be the least bad solution.

First of all, it is important to understand that the costs that the federal government is paying the utilities for continuing to store the spent fuel on site is small in comparison to the costs of reprocessing. As discussed in section II, DoE estimates that the costs will grow to $0.5 billion per year. We estimated the cost to be somewhat lower. Either cost is small, however, in comparison to a reprocessing program. Secretary of Energy Samuel Bodman has asked for an R&D budget ramping up to $0.8-0.9 billion per year in 2009 just to *assess* the cost of the GNEP program.[82] The French Government's figures for the extra cost of PUREX reprocessing LWR fuel and recycling the recovered plutonium once correspond to about $1 billion per year in the United States, and the National Academy of Science's estimate of the cost of a program involving sodium-cooled transuranic burner reactors was $1.6 to 3.2+ billion per year (1996 $).[83]

Secondly, it must be understood that interim storage of spent fuel would cost approximately the same if the federal government took possession of the spent fuel and moved it to a centralized storage site. The largest contribution to the cost of dry-cask storage is the storage casks. There would be economies of scale in the monitoring and maintenance costs at the centralized site, but these costs are quite modest for decentralized storage at sites with operating power plants because the casks require little maintenance and are stored within the plant's guarded perimeter. Any cost savings associated with centralized storage are likely to be offset by the fact that the infrastructure costs for dry-cask storage at the reactor sites will have already been paid for. There would also be the extra cost of transporting the spent fuel to the centralized storage site and then to Yucca Mountain or some other repository rather than transporting the spent fuel directly from the plant.[84]

Sometimes it is argued that continued storage of spent fuel at reactor sites creates a hazard. The amount of radioactivity that could be released from dry-cask storage is very small, however, in comparison to the potential releases from fuel in the reactor core or in a spent-fuel storage pool at operating reactor sites. The fuel in an operating reactor generates heat at a rate of about 30 kilowatts per kilogram. In a spent-fuel pool, a week after reactor shutdown, the fuel generates about 100 watts per kilogram. Loss of cooling water would result in the fuel in a reactor core heating up to combustion temperature within minutes. Recently discharged spent fuel in a pool would heat up to such temperatures within hours after a loss of water. Ten-year-old spent fuel generates about two watts of heat per kilogram and can be stored in dry casks passively cooled by air passing slowly over the outside surface of the canisters.[85] Air warmed by the radioactive decay heat rises and is replaced by cooler air. Even an attack with an anti-tank missile that breached a cask would release only a relatively small amount of radioactivity (see Figure 9).[86]

Two casks typically contain the equivalent of a year's spent fuel discharges from a 1,000 MWe nuclear power plant. Comparison of the simplicity of interim spent fuel storage with the complexity of the huge reprocessing complex shown in Figure 6 makes it easier to understand the relatively low cost of interim storage.[87]

Figure 9. Dry Cask Storage of Spent Fuel.

Why, then, are nuclear utilities in the United States pressing so hard for the government to begin moving the spent fuel off site? Perhaps one reason is that, in the 1970s, many nuclear-power opponents argued that there should be no further commitment to nuclear power until arrangements for ultimate disposal for spent fuel are in place. In 1976, in California, this became state law:

> no [new] nuclear fission thermal power plant . . . shall be permitted land use in the state . . . until both of the following conditions have been met:
>
> (a) The [California Energy] commission finds that there has been developed and that the United States through its authorized agency [the Nuclear Regulatory Commission] has approved and there exists a demonstrated technology or means for the disposal of high-level nuclear waste . . .[88]

The California law cannot be satisfied by the mere movement of spent fuel to a centralized storage site or to a reprocessing plant. The only way to satisfy it is through the licensing of a geological repository under Yucca Mountain or elsewhere.[89]

The position of the nuclear-power critics has evolved, however. In response to the Bush administration's reprocessing proposal, many groups that are critical of how nuclear power has been implemented in the United States have decided that they would prefer on-site dry-cask storage to reprocessing.[90]

On the other side of the debate, the Nuclear Energy Institute, which speaks for U.S. nuclear utilities, while acknowledging that the subsidies in the Energy Policy Act of 2005 for the first new nuclear power plants ordered since 1974 "clearly stimulated interest among electric utilities in constructing new nuclear power plants," insists that "[t]his increased interest requires [that] the federal government must meet its contractual responsibility to accept, transport, and dispose of used nuclear fuel through a comprehensive radioactive waste management program, including continued progress toward a federal used fuel repository."[91] Similarly, John Rowe, the President of Exelon, which manages 20 percent of U.S. nuclear capacity, has stated famously with regard to the urgency of licensing a federal waste repository, "We have to be able to look the public in the eye and say, 'If we build a plant, here's where the waste will go.' If we can't answer that question honestly to our neighbors, then we're playing politics too high for us to be playing."[92]

Note, however, that there is no requirement for reprocessing in the above statements of the nuclear-utility position. This suggests that the utilities might be willing to live with continued interim on-site storage as long as there is progress toward siting a repository.

The newly elected Senate Majority Leader, Harry Reid, who represents the State of Nevada, is, however, a dedicated opponent to the completion of the Yucca Mountain repository.[93] His proposed alternative is "The Spent Nuclear Fuel On-site Storage Act of 2005," which would have DoE take over responsibility for spent-fuel stored in dry casks at nuclear power plants to allow time for "a safe scientifically-based solution to be developed."[94]

VII. CONCLUSIONS

The U.S. Government's current interest in a federally-funded reprocessing program appears to be driven in significant part by an interest in finding a location to which it could ship the older spent fuel accumulating on power reactor sites. Shipments were to have begun to the Yucca Mountain geological repository in 1998, but the licensing of that repository has been delayed repeatedly and is now projected for 2017 at the earliest. If the federal government began to ship spent fuel to a reprocessing site, that would help it limit lawsuits by U.S. nuclear utilities that are seeking federal government reimbursement for their costs for prolonged on-site storage of spent fuel. The reprocessing option would be 4-8 times more costly, however, than on-site dry-cask storage for up to 50 years.[95]

At operating reactors, the incremental safety and security risk from such dry-cask storage of older fuel is negligible relative to the dangers from the fuel in the reactor core and the recently discharged hot fuel in the spent fuel pool.[96]

The nuclear-weapon proliferation costs of the United States unnecessarily embracing reprocessing as a necessary part of its nuclear fuel cycle cannot be quantified but could be severe.

*****

ACKNOWLEDGEMENTS

This report was originally proposed by Henry Sokolski, Director of the Nonproliferation Education Center, Washington, DC, and is posted on that organization's website at *www.npec-web.org* as well as on the website of the International Panel on Fissile Material (IPFM) *www.fissilematerials.org*.

The author gratefully acknowledges the generosity of Pierre Saverot in sharing information on the history of reprocessing and the cost of interim dry-cask spent-fuel storage.

*****

ABOUT THE IPFM

The International Panel on Fissile Materials (IPFM) was founded in January 2006. It is an independent group of arms-control and nonproliferation experts from both nuclear weapon and non-nuclear weapon states.

The mission of IPFM is to analyze the technical basis for practical and achievable policy initiatives to secure, consolidate, and reduce stockpiles of highly enriched uranium and plutonium. These fissile materials are the key ingredients in nuclear weapons, and their control is critical to nuclear weapons disarmament, to halting the proliferation of nuclear weapons, and to ensuring that terrorists do not acquire nuclear weapons. IPFM research and reports are shared with international organizations, national governments and nongovernmental groups.

The panel is co-chaired by Professor R. Rajaraman of the Jawaharlal Nehru University of New Delhi, India, and Professor Frank von Hippel of Princeton University. Its members include nuclear experts from 16 countries: Brazil, China, France, Germany, India, Japan, the Netherlands, Mexico, Norway, Pakistan, South Korea, Russia, South Africa, Sweden, the United Kingdom, and the United States.

Princeton University's Program on Science and Global Security provides administrative and research support for IPFM.

For further information about the panel, please contact the International Panel on Fissile Materials, Program on Science and Global Security, Princeton University, 221 Nassau Street, 2nd floor, Princeton, NJ 08542, or by email at *ipfm@fissilematerials.org*.

ENDNOTES - CHAPTER 7

1. President Ford, "Statement on Nuclear Policy," October 28, 1976: "I have concluded that the reprocessing and recycling of plutonium should not proceed unless there is sound reason to conclude that the world community can effectively overcome the associated risks of proliferation." President Carter, "Nuclear Power Policy," April 7, 1977: "[W]e will defer, indefinitely the commercial reprocessing and recycling of the plutonium produced in the U.S. nuclear power programs. From our own experience, we have concluded that a viable and economic nuclear power program can be sustained without such reprocessing and recycling," available from *www.nci.org/new/pu-repro/carter77a/index.htm*.

2. President Reagan, "Statement Announcing a Series of Policy Initiatives on Nuclear Energy," October 8, 1981: "I am lifting the indefinite ban which previous administrations placed on commercial reprocessing activities in the United States," available from *www.reagan.utexas.edu/archives/speeches/1981/100881b.htm*.

3. The first and only U.S. commercial reprocessing plant that operated was the Nuclear Fuel Services plant, at West Valley, New York. Between 1966 and 1972, about 640 tons of spent fuel was reprocessed there for a price of $33 per kilogram of heavy metal (kgHM) content, i.e., a total revenue of $21 million. The cost of satisfying new regulatory requirements and making major necessary plant modifications were estimated in 1976 at $600 million, which would have raised the reprocessing costs on new contracts to $350/kgHM (one-third the price that the French reprocessing company, Cogema, was charging at the time). U.S. utilities were unwilling to make additional contracts at the increased price, so the owner abandoned the plant, and it became a $4.5-billion federal cleanup site, *Nuclear Waste: Agreement Among Agencies Responsible for the West Valley Site is Critically Needed,*

GAO-01-314, 2001 and Pierre Saverot, personal communication, January 5, 2007.

4. *Nuclear Waste Policy Act of 1982*, Section 302, a5B.

5. *Nuclear Waste Policy Act* (as ammended in 1987), Section 160.

6. Statement of Edward Sproat III, Director, Office of Civilian Radioactive Waste Management, U.S. Department of Energy, to the Subcommittee on Energy and Air Quality of the House Committee on Energy and Commerce, September 13, 2006.

7. U.S. House of Representatives Appropriations Committee, *Report on the Energy and Water Development Appropriations Bill, 2006*, Report 109-86, May 18, 2005, p. 125.

8. There are 65 sites in the United States with operating power reactors. By 2011, 51 are expected to have dry storage, 57 by 2013, and 64 by 2017. Eight sites that no longer have operating reactors also have dry storage facilities, *Safety and Security of Commercial Spent Nuclear Fuel Storage*, Washington, DC: National Academies Press, 2006, pp. 20-24; *Going the Distance? The Safe Transport of Spent Nuclear Fuel and High-Level Radioactive Waste in the United States*, Washington, DC: National Academies Press, 2006, Table 5.2. By the time centralized storage could be established, therefore, almost all U.S. reactor sites will have installed dry storage systems, and the avoided costs would be the additional costs associated with capacity expansion. The cost for such expansion is approximately $110,000 per metric ton of uranium originally in the spent fuel, based on the published certificate of need filed January 18, 2005, by Xcel Energy with the Minnesota Public Utility Commission for a dry cask storage facility near the reactor building of its Monticello reactor. [The total estimated cost for this facility (p. 3-40) is $55 million, of which $2 million was for regulatory processes, $12 million for engineering and design, $4 million for plant upgrades (presumably to provide for transport of the spent fuel casks from the spent fuel pool to the storage site), $3.5 million for construction of the site for the dry cask storage, $26 million for 30 canisters and pre-fabricated storage modules, each of which will hold 61 fuel assemblies (approximately 10 tons of fuel), and $7.5 million for canister loading. The last two items, totaling $33.5 million or about $0.11 million per ton, are the incremental costs.] The annual cost for

operating and maintaining such a storage site is about $1 million per year at a site with operating reactors (Pierre Saverot, industry consultant, personal communication, October 29, 2006) and about $3.4 million per year at sites with shutdown reactors. [The latter number is an average of the $4.4 million per year ($1.9 million for staffing and security, $0.6 million for NRC (Nuclear Regulatory Commission) and state fees, $1.65 million in insurance and taxes, and $0.27 million for "other" costs) declared in the October 15, 2002, *Update of Site-Specific Decommissioning Costs* for the Maine Yankee nuclear power plant, Table 7-1, and the $2.4 million per year ($1.34 million for staffing and security, $0.33 million for NRC and state fees, and $0.7 million for insurance and taxes) declared by Pacific Gas and Electric in its July 27, 2004, response to a U.S. NRC request for supplemental Humboldt Bay Independent Spent Fuel Storage Installation financial information.] U.S. reactors discharge about 2,000 metric tons of spent fuel per year, *Safety and Security of Commercial Spent Nuclear Fuel Storage*, p. 20. At $0.11 million/ton, the annual incremental capital cost of dry cask storage for this amount of spent fuel would be about $220 million/year. At $1 million/year for 65 sites with operating reactors and $3.4 million/year at eight sites with shut down reactors, the cost of operating and maintaining the sites would be about $90 million/year. Overall, therefore, the cost estimate would be about $310 million/year. The DoE estimate may be higher in part because the it is being forced to pay infrastructure costs at the reactor sites as well as the incremental costs estimated here.

9. Section 114d of the *Nuclear Waste Policy Act* prohibits "the emplacement in the first repository of a quantity of spent fuel in excess of 70,000 metric tons of heavy metal or a quantity of solidified high-level radioactive waste resulting from the reprocessing of such a quantity of spent fuel until such time as a second repository is in operation." The radioactive waste associated primarily with the production of U.S. weapons plutonium is assumed to reduce the limit for civilian spent fuel to 63,000 tons.

10. Paul Lisowski, Deputy Program Manager, GNEP, presentation to the National Academies Nuclear and Radiation Studies Board, November 29, 2006, available from *www.ncsl.org/programs/environ/cleanup/NAS1106.htm*.

11. The findings of the Boston Consulting Group study, *Economic Assessment of Used Nuclear Fuel Management in the United States,* July 2006, which was commissioned by AREVA, will be discussed below.

12. House report on the *Energy and Water Development Appropriations Bill, 2007,* Report 109-474, May 19, 2006, pp. 64-65.

13. Based on *Plutonium Fuel: An Assessment,* Paris: OECD, 1989, Table 9.

14. Leo Szilard, "Memorandum on the Production of 94 [the atomic number of plutonium] and the Production of Power by the Means of the Fast Neutron Reaction," sent to A. H. Compton, January 8, 1943, cited in Bernard T. Feld, and Gertrud Weiss Szilard, eds., *The Collected Works of Leo Szilard: Scientific Papers,* Cambridge, MA: MIT Press, 1972, p. 178. The idea of this route to releasing the energy of U-238 first occurred to Szilard in May 1940 when he received a letter from Louis Turner, a Princeton physicist, who suggested that a new chain-reacting isotope might be formed as a result of neutron capture on U-238, William Lanouette, *Genius in the Shadows: The Biography of Leo Szilard, the Man Behind the Bomb,*New York: Charles Scribner's Sons, 1992, pp. 219-220.

15. If fully fissioned, 3 grams of uranium would release about 3 megawatt-days of heat or 260x109 joules. The combustion energy of a ton of coal is about 30x109 joules.

16. John Deutch, Ernest J. Moniz, Stephen Ansolabehere, Michael Driscoll, Paul E. Gray, John P. Holdren, Paul L. Joskow, Richard K. Lester, and Neil E. Todreas, *The Future of Nuclear Power: An Interdisciplinary MIT Study,* Cambridge, MA: Massachusetts Institute of Technology, July 2003.

17. International Atomic Energy Agency, *Power Reactor Information System (PRIS) Database,* updated January 5, 2006, available from *www.iaea.org*.

18. A list of 11 shutdown and 8 operational fast-neutron reactors as of 1995 is given in David Albright, Frans Berkhout, and William Walker, *Plutonium and Highly Enriched Uranium*

1996, Oxford, UK: Oxford University Press, 1997, p. 196. Since that time, two more reactors (Kazakhstan's BN-350 and France's Superphénix) have been shut down permanently, one (Japan's Monju) has been shut down for more than a decade by a sodium fire, and France's Phénix is scheduled to be shut down. Russia's BN-600 has been kept on line with an average capacity factor of about 74 percent since 1980 but has had 15 sodium fires in 23 years, N. N. Oshkanov, M. V. Bakanov, and O. A. Potapov, "Experience in Operating the BN-600 Unit at the Belyiyar Nuclear Power Plant," *Atomic Energy,* Vol. 96, No. 5, 2004, p. 315. Three new sodium-cooled reactors are being built in: China (25 Mwe {megawatts electric]), India (500 MWe), and Russia (800 MWe).

19. Matthew Bunn, Steve Fetter, John Holdren and Bob van der Zwaan, "The Economics of Reprocessing Versus Direct Disposal of Spent Nuclear Fuel," *Nuclear Technology,* Vol. 150, June 2005, p. 209, updated by Steve Fetter. The price of uranium on the spot market has increased further since 2005, to about $350/kg in mid-2007 before declining to $230/kg at the end of 2007, *Nukem Market Report,* December 2006. This appears to be due to the closure of many mines during the period of low prices and has stimulated renewed interest in uranium mining. The spot market price peak therefore is likely to be as transient as the peak in the late 1970s, "Uranium Glowing," *Economist,* August 19, 2006, p. 53.

20. Richard C. Hewlett and Francis Duncan, *Nuclear Navy 1946-1962,* Chicago, IL: University of Chicago Press, 1974, p. 274. I would like to thank Thomas B. Cochran for bringing this quotation to my attention.

21. Belgium, France, Germany, India, Japan, Russia, the United Kingdom, and the United States, Albright, *et al., Plutonium and Highly Enriched Uranium 1996,* p. 156. China has recently completed but not yet operated a pilot reprocessing plant.

22. Bunn, *et al.,* "The Economics of Reprocessing Versus Direct Disposal of Spent Nuclear Fuel."

23. Shaun Burnie and Aileen Mioko Smith, "Japan's Nuclear Twilight Zone," *Bulletin of the Atomic Scientists,* May/June 2001, p. 58.

24. "Summary," *Management of Separated Plutonium,* London, UK: The Royal Society, 1998 .

25. This is an update of Table 2.A in *Global Fissile Material Report 2006,* available from *www.fissilematerials.org*.

26. "Sellafield Shutdown Ends the Nuclear Dream," *The Guardian,* August 26, 2003; "Huge Radioactive Leak Closes THORP Nuclear Plant," *The Guardian*, May 9, 2005.

27. The German utilities began to build their own domestic reprocessing plant but encountered massive public opposition and found also that it would be cheaper to accept offers to invest in the French and British plants. An excellent history of this episode has been written by Pierre Saverot, *History of the Wackersdorf Reprocessing Plant Project*, Fairfax, VA: JAI Corporation, JAI-557, 2003.

28. Mark Hibbs, "German At-Reactor Storage Deal May Scuttle Gorleben, Ahaus," *Nuclear Fuel*, February 8, 1999; and Ann MacLachlan, "Cogema Says German Phase-Out Pact 'No Surprise,' Doesn't Change Situation," *Nuclear Fuel*, June 26, 2000.

29. Tadahiro Katsuta and Tatsujiro Suzuki, *Japan's Spent Fuel and Plutonium Management Challenges,* International Panel on Fissile Materials, 2006.

30. *Conference Report on the Energy and Water Appropriations Act for Fiscal Year 2006*, Report 109-275, "Nuclear Energy Programs," pp. 141-142; and "Nuclear Waste Disposal," pp. 156-157.

31. *Report to Congress: Spent Nuclear Fuel Recycling Program Plan,* Washington, DC: U.S. Department of Energy, May 2006, Figure 3.

32. Ronald Wigeland, Theodore Bauer, Thomas Fanning, and Edgar Morris, "Separations and Transmutation Criteria to Improve Utilization of a Geological Repository," *Nuclear Technology,* April 2006, p. 95.

33. Proposed "Nuclear Fuel Management and Disposal Act," submitted to the Congress, April 5, 2006.

34. J. Kessler, *Room at the Mountain: Analysis of the Maximum Disposal Capacity for Commercial Spent Fuel in a Yucca Mountain Repository,* Palo Alto, CA: Electric Power Research Institute, Report 1013523, May 2006.

35. Office of Nuclear Energy, Department of Energy, "Notice of Request for Expressions of Interest in a Consolidated Fuel Treatment Center to Support the Global Nuclear Energy Partnership," and "Notice of Request for Expressions of Interest in an Advanced Burner Reactor to Support the Global Nuclear Energy Partnership," *Federal Register,* Vol. 71, No. 151, August 7, 2006, pp. 44673-44679.

36. *Global Nuclear Energy Partnership Strategic Plan,* U.S. Department of Energy, GNEP-167312, January 2007, p. 9; "Department of Energy: Notice of Intent to Prepare a Programmatic Environmental Impact Statement for the Global Nuclear Energy Partnership," *Federal Register,* Vol. 72, No. 2, January 4, 2007, pp. 331-336.

37. Assuming a burn up (fission energy release) of 50 MWt [megawatt thermal]-days per kilogram of uranium in the fuel. Fuel discharged in the 1970s had lower burn up, and the plutonium-241 has decayed into americium-241. Its plutonium content is therefore about 0.8 percent, *Plutonium Fuel: An Assessment,* Table 9.

38. See e.g., *International Fuel Cycle Evaluation: Fast Breeders,* Vienna, Austria: International Atomic Energy Agency, INFCE/PC/5, 1980, p. 2.

39. *Nuclear Wastes: Technologies for Separations and Transmutation,* Washington, DC: National Academies Press, 1996.

40. *Economic Assessment of Used Nuclear Fuel Management in the United States,* Boston, MA: Boston Consulting Group, July 2006.

41. *Nuclear Wastes: Technologies for Separations and Transmutation,* p. 7.

42. See *Nuclear Waste Fund Fee Adequacy: An Assessment*, Washington, DC: Department of Energy, DoE/RW-0534, May 2001.

43. See, for example, the testimony of David J. Modeen, Vice President and Chief Nuclear Officer, Electric Power Research Institute, before the Energy Subcommittee of the Science Committee, U.S. House of Representatives, April 6, 2006.

44. *2002 Waste Stream Projection Report*, TDR-CRW-SE-000022, Rev. 1, DoE, 2003.

45. The number of fast-neutron reactors that would be required depends upon their net destruction rate of transuranics. Assuming a 40 percent thermal-to-electric conversion efficiency, a 1000 MWe fast-neutron reactor operating at a 90 percent capacity factor would fission approximately 0.8 tons of transuranics per year. If the reactor did not create new transuranics at the same time, fissioning the 24 tons of transuranics in the spent fuel discharged annually by 100 GWe of U.S. LWRs, would require 30 such reactors. Even with their uranium blankets removed, however, fast-neutron reactors, as currently conceived, would produce transuranics. The National Academy of Sciences' report discussed the tradeoffs in reducing this conversion ratio below unity as follows:

> With reduced fissile breeding and reduced heavy metal inventory, the burner designs also result in increased reactivity swing over a fuel cycle. This then requires larger control rod worths and hence entails potentially larger positive reactivity insertions [if those rods are withdrawn] and degraded performance in transient overpower events. As the breeding ratio decreases, there is less reactivity insertion resulting from sodium voiding in a power excursion. However, with decreasing breeding ratio, less negative reactivity is available from Doppler broadening of the neutron absorption resonances that occurs when the fuel is heated in a power excursion. Based on these considerations, GE concludes that a TRU [transuranium] burner with a breeding ratio of 0.6 and a core height of 0.76 m is the lowest possible breeding ratio configuration that would have acceptable safety features.

Nuclear Wastes: Technologies for Separations and Transmutation, pp. 205-6. For a conversion ratio of 0.6, 75 GWe of burner-reactor capacity would be required. Fast-neutron reactor advocates at Argonne National Lab argue, however, that a conversion ratio as low as 0.25 can be achieved safely with added control rods and twice-annual refueling, J. E. Cahalan, M. A. Smith, R. N. Hill, and F. E. Dunn, "Physics and Safety Studies of a Low Conversion Ratio Sodium Cooled Fast Reactor," *Proceedings of the PHYSOR 2004 Conf.*, Chicago, IL: American Nuclear Society, April 2004. For a 0.25 conversion ratio, 40 GWe of burner-reactor capacity would be required.

46. The long-term evacuation caused by the Chernobyl accident is primarily due to land contamination by cesium-137. "Exposures and Effects of the Chernobyl Accident," *Sources and Effects of Ionizing Radiation*, United Nations Scientific Committee on the Effects of Atomic Radiation, U.N., 2000, Annex J.

47. *Economic Assessment of Used Nuclear Fuel Management in the United States.*

48. The maximum amount of spent fuel that AREVA has reprocessed in its reprocessing complex in one year was 1650 tons in 1996; between 1996 and 2004, it reprocessed an average of about 1300 tons per year. Mycle Schneider, personal communication, March 1, 2006. AREVA's MELOX plant is licensed to produce up to 195 tons of MOX fuel per year. That much fuel would contain about 14 tons of power-reactor plutonium.

49. Image courtesy of Greenpeace (image 35044).

50. *Report to Congress: Disposition of Surplus Defense Plutonium at Savannah River*, Washington, DC: DoE-National Nuclear Security Administration, February 15, 2002, p. 4-7.

51. MOX containing weapon-grade plutonium would contain only about 4 percent plutonium versus 7 percent for fuel containing reactor-grade plutonium.

52. DoE Inspector General Audit Report, *The Status of the Mixed Oxide Fuel Fabrication Plant*, 2005.

53. See, e.g., *Department of Energy: Opportunity to Improve Management of Major System Acquisitions,* Washington, DC: U.S. Government Accountability Office, GAO/RCED-97-17, 1996; and *Department of Energy: Further Actions Needed to Strengthen Contract Management for Major Projects,* GAO)-05-123, 2005.

54. *Economic Assessment of Used Nuclear Fuel Management in the United States,* Figure 36.

55. *Ibid.*, Appendix 5, Figure 29, and p. 31.

56. *Ibid.*, Appendix 10.

57. J. M. Charpin, B. Dessus and R. Pellat, *Report to the Prime Minister [of France]: Economic Forecast Study of the Nuclear Power Option,* 2000, p. 38, says 150 years. The June 2001 report of France's National Assessment Committee on Radioactive Waste Management Research says "centuries," Pierre Saverot, personal communication, January 4, 2007.

58. In equilibrium, about 12 of the approximately 100 GWe of U.S. LWR capacity would be devoted to MOX recycle. This would leave 88 GWe of capacity fueled by LEU. The basis for the fast-reactor to LEU-fueled LWR ratio is explained in a footnote to the subsection on the 1996 National Academy of Sciences study above. The AREVA study mentions the option of recycling once in LWRs and then in fast reactors, but provides no basis for its estimate of the number of fast reactors required: "we assume that ~ 10-15 reactors are needed to absorb the plutonium stock," p. 78.

59. *Economic Assessment of Used Nuclear Fuel Management in the United States,* p. 78.

60. The capital cost of the 1.2 GWe Superphénix in French Francs was FF 34.4 billion (about $7 billion in 1996$) according the France's public accounting tribunal, the Cour des Comptes, "Accounting Panel Pegs Superphénix [construction plus decommissioning] Cost at FF 60-Billion to 2000," *Nucleonics Week,* October 17, 1996.

61. Charpin, Dessus, and Pellat, *Report to the Prime Minister: Economic Forecast Study of the Nuclear Power Option*. We have taken a 1999 French Franc (FF) = $0.2 (2006).

62. *Report to Congress: Spent Nuclear Fuel Recycling Program Plan*, p. 10.

63. Thomas B. Cochran, William M. Arkin, Robert S. Norris, and Milton M. Hoenig, *U.S. Nuclear Warhead Facility Profiles*, Cambridge, MA: Ballinger Press, 1987, p. 24.

64. *National Energy Policy*, Washington, DC: The White House, 2001, p. 5-17.

65. We use here the nomenclature for the variants of UREX+ used by E. D. Collins in "Closing the Fuel Cycle Can Extend the Lifetime of the High-Level-Waste Repository," Washington, DC" American Nuclear Society, 2005 Winter Meeting, November 17, 2005.

66. G. F. Vandergrift, "Designing and Demonstration of the UREX+ Process Using Spent Nuclear Fuel," presentation at the International Conference on Advances for Future Nuclear Fuel Cycles, Nîmes, France, June 21-24, 2004.

67. The IAEA's threshold for self protection is one Sievert per hour at a distance of 1 meter, "The Physical Protection of Nuclear Material and Nuclear Facilities," INFCIRC/225, Rev. 4. A dose of several Sieverts would be lethal.

68. Collins, "Closing the Fuel Cycle Can Extend the Lifetime of the High-Level-Waste Repository."

69. J. Kang and F. von Hippel, "Limited Proliferation Resistance Benefits from Recycling Unseparated Transuranics and Lanthanides from Light-Water Reactor Spent Fuel," *Science & Global Security*, Vol. 13, 2005, p. 169.

70. Rube Goldberg was a cartoonist who specialized in designing enormously complex systems to accomplish simple tasks, available from *www.rube-goldberg.com/*.

71. Phillip Finck, Deputy Associate Laboratory Director for Applied Science and Technology and National Security, Argonne National Laboratory, "The Benefits of the Closed Nuclear Fuel Cycle," briefing to U.S. House of Representatives staff, March 10, 2006, slide 3.

72. This section was stimulated by Edwin Lyman's report, "The Global Nuclear Energy Partnership: Will it Advance Nonproliferation or Undermine It?" *Proceedings of the Annual Meeting of the Institute for Nuclear Materials Management*, Nashville, TN, July 16-20, 2006.

73. Marvin M. Miller, "Are IAEA Safeguards on Plutonium Bulk-Handling Facilities Effective?" Washington, DC: Nuclear Control Institute, 1990.

74. Kang and von Hippel, "Limited Proliferation Resistance Benefits from Recycling Unseparated Transuranics and Lanthanides from Light-Water Reactor Spent Fuel."

75. DoE, "Notice of Request for Expressions of Interest in a Consolidated Fuel Treatment Center to Support the Global Nuclear Energy Partnership."

76. Boston Consulting Group, *Economic Assessment of Used Nuclear Fuel Management in the United States.*

77. "Fact Sheet: Strengthening International Efforts Against WMD Proliferation," Washington, DC: The The White House, February 11, 2004, available from *www.whitehouse.gov/news/releases/2004/02/20040211-5.html.*

78. *Multilateral Approaches to the Nuclear Fuel Cycle: Expert Group Report Submitted to the Director General of the International Atomic Energy Agency,* IAEA, INFCIRC-640, 2005, pp. 102-103, 119.

79. Argentina and Brazil have already been informed by the Bush administration that they are exempt from the proposed stricture, U.S. State Department officials, personal communication. It seems likely that Australia and Canada will be as well. South

Africa and Ukraine have also expressed an interest in national enrichment plants.

80. Mark Hibbs, "US Has not Decided Whether GNEP Will Include ROK Pyroprocessing," *Nuclear Fuel,* September 11, 2006.

81. "AREVA dual-track strategy aimed at two reprocessing plants," *Nuclear Fuel,* July 3, 2006. The countries listed as possible buyers were Japan, Germany, Canada, the Netherlands, Australia, Brazil, Argentina, and South Africa.

82. "Energy Secretary Bodman Outlines Plans on Yucca, Nuclear Waste, and Oil Security," Washington, DC: Environment and Energy Publishing, March 8, 2006.

83. The French Government's estimate was $33 billion extra for 58,000 tons of spent fuel (see Table 2). The National Academy of Sciences' report, *Nuclear Wastes,* estimated an extra cost of $50-100+ billion for 62,000 tons of spent fuel. We have converted these to annual costs by using the current U.S. spent-fuel discharge rate of 2,000 tons per year.

84. Allison Macfarlane, "Interim Storage of Spent Fuel in the United States," *Annual Review of Energy and Environment,* Vol. 26, 2001, p. 201.

85. Robert Alvarez, Jan Beyea, Klaus Janberg, Jungmin Kang, Ed Lyman, Allison Macfarlane, Gordon Thompson, and Frank von Hippel, "Reducing the Hazards from Stored Spent Power-Reactor Fuel in the United States," *Science & Global Security,* Vol. 11, 2003, p. 1.

86. *Safety and Security of Commercial Spent Nuclear Fuel Storage,* pp. 68-69.

87. Picture of the spent fuel storage area of the shut down Maine Yankee nuclear power plant, available from *www.maineyankee.com/.*

88. California Public Resources Code, para. 25524.2.

89. California Energy Commission, *Integrated Energy Policy Report,* 2005 p. 85.

90. As of November 2, 2006, 21 national groups and 106 local groups had signed a position statement, "Principles for Safeguarding Nuclear Waste at Reactors," available from *www.citizen.org/documents/PrinciplesSafeguardingIrradiatedFuel.pdf.* The statement indicates that the groups would require in addition that spent-fuel pools and dry-cask storage be hardened against attack and that the spent-fuel be moved to dry cask storage after 5 years in storage pools. Both requirements could be met at costs modest relative to reprocessing.

91. Skip Bowman, President of the Nuclear Energy Institute, testimony before the Senate Energy and Natural Resources Committee, May 16, 2006.

92. Cora Daniels, "Meet Mr. Nuke: John Rowe of Exelon is Staking Out a Megashare of What Could Be America's Great Atomic Revival," *Fortune,* May 15, 2006.

93. "The Proposed Yucca Mountain Nuclear Waste Dump Is Never Going To Open," available from *www.reid.senate.gov/issues/yucca.cfm.*

94. *Ibid.*

95. At the $50-100+ billion estimated in the 1996 National Academy of Sciences study, reprocessing of 62,000 tons spent LWR fuel and fission of the recovered transuranics in advanced burner reactors would cost $0.8-1.6+ million per metric ton. For comparison for a site with operating reactors and 500 tons of dry cask fuel, the incremental capital cost for dry-cask storage is $0.11 million/ton, and the annual cost would be about $0.002 million per ton-year. For a similar site with shut down reactors, the annual cost would be about $0.009 million per ton-year. See footnote to section II for details.

96. As of the end of 2002, less than 3,000 out of 50,000 tons of U.S. spent fuel were stored at 10 U.S. sites with no operating power reactors, *Going the Distance,* Table 2.

APPENDIX

SCENARIOS FOR THE FRENCH FUEL CYCLE*

	Percentage of Spent LEU Fuel Reprocessed			
	% (S6)	27% (Reprocessing Ends in 2010, S4)	100% (Derived Scenario)	No Reprocessing (S7)
	Fuel cycle costs (10^9 1999 FF [2006 $] undiscounted)			
Front end	578 [116]	602 [120]	558 [112]	611 [122]
Back end	370 [74]	307 [61]	422 [84]	203 [41]
Net	948 [190]	909 [182]	980 [196]	814 [162]
Back end cost ($/kg)			$1450	$700
Back end cost ($$10^{-3}$/kWh)			4.2	2.0
	Inputs			
Natural uranium mined (10^3 metric tons)	437	460	418	475
Separative Work (million SWUs)	313	330	299	341
LEU fuel fabricated (10^3 tons uranium)	54	56	52	58
MOX fuel fabricated (10^3 tons)	4.8	2	7.1	0
LEU fuel reprocessed (10^3 tons)	36	15	52	0
	Wastes			
Depleted uranium (10^3 tons)	379	401	360	417
LEU Spent fuel (10^3 tons)	18	41	0	58
MOX Spent Fuel (10^3 tons)	4.8	2	7.1	0
Transuranic Waste (10^3 cubic meters)	18	12	23	0
High-level waste (10^3 cubic meters)	4.8	1.6	7.5	0
Plutonium/Americium in spent fuel (tons)	514	602	441	667
Reprocessed uranium (10^3 tons)	34	14	50	0

*Assuming a 45-year average life for France's LWR fleet. In all scenarios, 20.2×10^{12} kilowatt hours are generated, J.M. Charpin, B. Dessus and R. Pellat, *Report to the Prime Minister: Economic Forecast Study of the Nuclear Power Option*, 2000, Tables on pp. 43, 56, 214., 215. We assumed that a 1999 French Franc (FF) = $0.2 (2006$).

CHAPTER 8

THE NPT, IAEA SAFEGUARDS AND PEACEFUL NUCLEAR ENERGY: AN "INALIENABLE RIGHT," BUT PRECISELY TO WHAT?

Robert Zarate

In mid-October 2006, a few days after North Korea's surprise detonation of a nuclear explosive device, the Director General of the International Atomic Energy Agency (IAEA) sounded the alarm on what he now sees as a troubling trend: the growing number of states seeking to enrich uranium, reprocess spent nuclear fuel to separate from it plutonium, and engage in other sensitive nuclear fuel-making activities that provide direct access to weapons-ready fissile material.[1] During an address to the IAEA's symposium on international safeguards, Dr. Mohamed ElBaradei candidly acknowledged that nuclear fuel-making "creates many new challenges, both for the international community and for [the Agency], because verifying enrichment facilities or reprocessing facilities is quite difficult, and the so-called conversion time"—that is, the time required to convert fissile material for use in a nuclear explosive device—"is very short."[2] Then, the IAEA Director General went so far as to say that when non-nuclear-weapon states become nuclear fuel-makers, then "we are dealing with what I call *virtual nuclear-weapon states*."[3]

As North Korea's recent nuclear detonation and Iran's ongoing nuclear intransigence demonstrate, the emergence of more nuclear fuel-making states—of what ElBaradei now describes as *virtual nuclear-weapon states*—not only challenges the continuing relevance of the *Treaty on the Nonproliferation of Nuclear Weapons* (Nuclear Nonproliferation Treaty or NPT)[4] and the IAEA safeguards system, but also threatens the security of the many nuclear-weapon states and non-nuclear-weapon states that participate in the NPT-IAEA safeguards system. For if a non-nuclear-weapon state has acquired fuel-making capabilities sufficient to accumulate stocks of fissile material (principally in the form of highly enriched uranium or separated plutonium), then that state has cleared *the* most difficult obstacle on the path to its first nuclear explosive. This is why, during the May 2005 quadrennial NPT review conference, then-Secretary-General of the United Nations Kofi Annan called attention to what he called the "Janus-like character" of nuclear fuel-making:

> The [nonproliferation] regime will not be sustainable if scores more States develop the most sensitive phases of the fuel cycle and are equipped with the technology to produce nuclear weapons on short notice—and, of course, each individual State which does this only will leave others to feel that they must do the same. This would increase all the risks—of nuclear accident, of trafficking, of terrorist use, and of use by States themselves.[5]

The extent to which the Nuclear Nonproliferation Treaty precludes—or should be interpreted as precluding—"the most sensitive phases of the fuel cycle" remains unclear, however. On the one hand, the NPT's Articles I and II articulate the fundamental, corresponding, and overriding responsibilities of the legally-recognized nuclear-weapon signatories and non-nuclear-weapon signatories, the *sine qua non* obligations that make this treaty a *non*proliferation treaty.[6] On the other hand, the NPT's Article IV recognizes both the "inalienable right" of signatories "to develop research, use, and production of nuclear energy for peaceful purposes without discrimination and in conformity with articles I and II"; and the right of signatories "to participate" in the "fullest possible exchange of equipment, materials, and scientific and technological information for the peaceful uses of nuclear energy."[7] Precisely what Articles I and II should prohibit, and when and how these prohibitions should apply to Article IV and the most weapons-relevant civilian applications of nuclear technology, continue to be a matter of heated debate.

Article IV never explicitly mentions enrichment, reprocessing, and other nuclear fuel-making technologies, yet some governments nevertheless interpret Article IV as implicitly recognizing the specific or per se right of signatories to any nuclear technological activities that can be conceivably labeled "peaceful," short of actually inserting fissile material into a nuclear explosive device.[8] Under this interpretation, all that is required is that a non-nuclear-weapon signatory conclude, in accordance with the NPT's Article III, a comprehensive safeguards agreement with the IAEA; that the IAEA administer safeguards on the nuclear materials involved in the civilian nuclear activities of the signatory; and that the signatory be in full compliance with its NPT and IAEA safeguards obligations.

The Islamic Republic of Iran has pushed the per se right interpretation of Article IV much, much further, however. In 2003, the IAEA became aware of the broad range of sensitive nuclear materials and technologies that the Iranian government had concealed from it for nearly two decades.[9] Over the next two years, Tehran failed to cooperate fully and transparently as IAEA inspectors attempted to reconstruct the shrouded history of the Islamic Republic's nuclear program and ensure the absence of undeclared nuclear activities in Iranian territory in order to verify both the correctness and completeness of Iran's declarations to the Agency.[10] In September 2005, the IAEA Board of Governors responded to Tehran's lack of cooperation by finding Iran to be in non-compliance with its NPT and IAEA safeguards obligations.[11] As a consequence, the IAEA Board declared—and subsequently the Security Council of the United Nations decided in a legally-binding manner—that Iran should suspend all nuclear fuel-making activities *until* the IAEA fully resolves the many serious issues surrounding Iran's history of non-compliance.[12] In rejecting any suspension, though, Iranian officials have argued that absolutely no circumstance whatsoever—not even a finding of non-compliance by the IAEA Board or a legally-binding resolution from the UN Security Council—can limit what they interpret to be their government's "specific and undeniable right" to enrichment, reprocessing, and other sensitive nuclear fuel-making activities under the NPT.[13] In short, the Iranian government claims that Article IV recognizes not merely the per se right, but rather the per se right without any qualification whatsoever, of signatories to nuclear fuel-making.

Given the many challenges that the spread of enrichment, reprocessing, and other sensitive nuclear fuel-making technologies pose to the NPT-IAEA safeguards system, this chapter addresses two related questions:

1. To what extent can the IAEA, given its own safeguarding goals, effectively safeguard nuclear materials—especially weapons-ready nuclear materials involved in nuclear fuel-making and other sensitive activities?

2. Does the NPT recognize the right of signatories to develop, access, or use nuclear materials and technologies that the IAEA cannot effectively safeguard, even if these unsafeguardable materials and technologies are claimed to be "for peaceful purposes"?

Answers to these questions have far-reaching implications for the nuclear energy and nonproliferation policies of individual governments and international organizations. Indeed, these questions go to the very heart of the global nonproliferation system's rationale for sharing widely the civilian uses of nuclear technology. If, in fact, the IAEA actually is capable of safeguarding effectively even the most sensitive nuclear materials and technologies, then this provides a strong warrant for interpretations that view the NPT's Article IV as permitting any and all nuclear activities short of inserting fissile material into a nuclear weapon. However, if the IAEA cannot safeguard effectively all nuclear materials and technologies, then broadly permissive interpretations of Article IV become not only unwarranted, but also perversely detrimental to the NPT's fundamental goal of nuclear *non*proliferation.

With respect to the first question, this chapter argues that the IAEA, given its own safeguarding criteria, remains unable to safeguard effectively a broad range of sensitive nuclear materials, technologies, and activities. In particular, the Agency cannot provide—even in principle—timely warning of a non-nuclear-weapon state's diversion of weapons-ready nuclear materials from civilian applications to nuclear weapons or unknown purposes; it must tolerate, under its current accounting methods, large amounts of unaccounted nuclear material at facilities that handle such material in bulk form before even beginning to suspect a diversion; and it appears to lack adequate financial resources to carry out many of its safeguarding activities effectively.

With respect to the second question, this chapter argues that—in conformity with the generally accepted principles of treaty interpretation that the *Vienna Convention of the Law of Treaties* codifies—the NPT, at a minimum, can be interpreted as not recognizing the "inalienable right" of signatories to nuclear materials, technologies, and activities that the IAEA cannot effectively safeguard. The NPT's Article IV appears to establish three legally-binding qualifications that clarify the scope of the "nuclear energy for peaceful purposes" to which signatories have a "right" to develop and use—a key qualification being the effective safeguardability of civilian application of nuclear technology and related nuclear materials. But while the NPT may be understood as prohibiting non-nuclear-weapon signatories from unsafeguardable nuclear materials, technologies, and activities, the treaty also provides for mechanisms by which nuclear-weapon signatories can provide, individually or through multilateral frameworks, non-nuclear-weapon signatories with the benefits of proscribed, unsafeguardable peaceful applications of nuclear technology in an economically-sound, nondiscriminatory manner. In short, though some governments continue to insist on reading the NPT as implying the per se right—and now in Iran's case, the unqualified per se right—of signatories to nuclear fuel-making, Article IV need not be interpreted as providing *de jure* cover for the *de facto* status of a *virtual nuclear-weapon state*. Indeed, the NPT can be read in a more sustainable way.[14]

The remainder of this chapter proceeds in three sections. The first section revisits key IAEA safeguards documents in order to unpack the Agency's goals for effective safeguarding, and analyzes the extent to which the Agency actually can meet these goals when it administers safeguards. The second section uses the generally accepted principles of treaty interpretation, as codified by the *Vienna Convention of the Law of Treaties*, to explore the extent to which the NPT recognizes the "inalienable right" of signatories to, and prohibits signatories from, sensitive nuclear materials, technologies, and activities—especially those that the IAEA cannot effectively safeguard. Finally, the conclusion considers what the chapter's analysis on the NPT and IAEA safeguards system implies for national and multilateral policies to limit and manage the dangers of nuclear proliferation.

ASSESSING THE EFFECTIVENESS OF IAEA SAFEGUARDS

The Nuclear Nonproliferation Treaty's preamble stresses not only the importance of "the principle of *safeguarding effectively* [emphasis added] the flow of source and special fissionable materials by use of instruments and other techniques at certain strategic points," but also the need for "research, development, and other efforts to further the [principle's] application, within the framework of the International Atomic Energy Agency safeguards system . . . ."[15] The extent to which the IAEA today can actually administer effective safeguards on nuclear materials—especially weapons-ready nuclear materials involved in civilian applications of nuclear technology—remains unclear, however.

In accordance with the NPT preamble's principle of effective safeguarding, Article III requires that each non-nuclear-weapon signatory conclude a comprehensive safeguards agreement with the IAEA "for the exclusive purpose of verification of the fulfillment of [the signatory's] obligations assumed under this Treaty with a view to preventing the diversion of nuclear energy from peaceful purposes to nuclear weapons and other nuclear explosive devices."[16] In 1972, 2 years after the NPT entered into force, the IAEA released *The Structure and Content of Agreements Between the Agency and States Required in Connection with the Treaty on the Non-Proliferation of Nuclear Weapons* ("Model Comprehensive Safeguards Agreement" or "INFCIRC/153"), which defines the technical objective of safeguards as "the timely detection of diversion of significant quantities of *nuclear material* from peaceful nuclear activities to the manufacture of nuclear weapons or of other nuclear explosive devices or for purposes unknown, and deterrence of such diversion by the risk of early detection" (emphasis in the original).[17] To meet this objective, the *Model Comprehensive Safeguards Agreement* identifies the IAEA's means as "the use of material accountancy as a safeguards measure of fundamental importance, with containment and surveillance as important complementary measures."[18]

When the Agency released INFCIRC/153, it had not yet determined the specific methods and metrics to evaluate the effectiveness of safeguards. In the mid-to-late 1970s, however, the IAEA's Standing Advisory Group on Safeguards Implementation ("SAGSI") used the numerical estimates of four terms from INFCIRC/153—namely, *significant quantity, timely detection, risk of detection,* and *probability of raising a false alarm*—to define precisely the Agency's "detection goals" (emphasis added).[19] In theory, these detection goals provide the IAEA with ways to measure the extent to which it is obtaining INFCIRC/153's safeguards objective, and verifying the fulfillment of NPT-signatory obligations. In practice, though, the Agency cannot meet these goals with respect to a wide range of nuclear materials and civilian applications of nuclear fuel-making technology. Although it is far beyond the scope of this chapter to describe in exhaustive detail every difficulty that the IAEA faces in attempting to safeguard effectively, the sections below summarize representative examples of these difficulties.

Abrupt Diversion of Nuclear Materials: Conversion Time vs. Timely Warning.

When the IAEA administers safeguards, it aims to account for and inspect declared nuclear materials in civilian applications of nuclear technology frequently enough to detect the diversion of a significant quantity ("SQ") of nuclear material before it has been—or can be—converted into a bomb. A *significant quantity* is defined by the Agency as "the approximate amount of nuclear material for which the possibility of manufacturing a nuclear explosive device cannot be excluded."[20] Table 1 gives the SQ values that the IAEA currently uses. Here, it is worth noting that some analysts have concluded that the IAEA's current SQ values are inadequate. For example, in October 2005 Thomas Cochran of the National Resources Defense Council argued that "the IAEA's SQ values for direct use materials are not technically valid or defensible," and that, in some circumstances, "the SQ values for direct use plutonium and high enriched uranium (HEU) [should] be reduced by a factor of about eight."[21]

Direct-Use Material	**SQ**
Plutonium (containing < 80% ^{238}Pu)	8 kg Pu
Uranium-233	8 kg ^{233}U
High Enriched Uranium ($^{235}U \geq 20\%$)	25 kg ^{235}U
Indirect-Use Material	**SQ**
Uranium ($^{235}U < 20\%$)[a]	75 kg ^{235}U (or 10 t natural U or 20 t depleted U)
Thorium	20 t Th
a. Including low enriched, natural and depleted uranium.	
Data Source: *IAEA Safeguards Glossary*, 2001 Ed., International Nuclear Verification Series No. 3, Vienna, Austria: IAEA, June 2002, sec. 3, para. 13, Table II.	

Table 1. IAEA's Estimated Values for Significant Quantities.

To express quantitatively the extent to which a non-nuclear-weapon state in possession of at least one SQ of diverted nuclear material could pose an immediate proliferation threat, the IAEA uses a metric known as *conversion time*, defined as "the time required to convert different forms of nuclear material to the metallic components of a nuclear explosive device."[22] In order to provide *timely warning* of a non-nuclear-weapon state's diversion of nuclear material "from peaceful nuclear activities to the manufacture of nuclear weapons or of other nuclear explosive devices or for purposes unknown" *so that* governments can organize diplomatic and other forms of pressure on the diverting state, the numerical value of the IAEA's *timeliness detection goal* for a given category of nuclear material should be, in principle, much less than the value of its estimated *conversion time* for that category of nuclear material.[23]

Even in principle, though, this is not always the case. Table 2 compares the IAEA's estimated *conversion time* for special and source nuclear materials with its corresponding *timeliness detection goals* in states where either the IAEA's Additional Protocol (a voluntary agreement which grants the Agency greater inspection authority) has not entered into force; or the Agency has not concluded the absence of undeclared nuclear material or activities, and thus has not verified the completeness of the state's declarations.

Type Nuclear Material	Est. Conversion Time	Timeliness Detection Goal
Unirradiated Direct-Use (Metallic Form)[a]	7 – 10 days	1 month
Unirradiated Direct-Use (Chemical Compounds/ Mixtures)[b]	7 – 21 days	1 month
Irradiated Direct-Use[c]	1 – 3 months	3 months
Indirect Use[d]	3 – 12 months	12 months

a. Pu, HEU or ^{233}U metal.
b. PuO_2, $Pu(NO_3)_4$ or other pure Pu compounds; HEU or ^{233}U oxide or other pure U compounds; MOX or other non-irradiated pure mixtures containing Pu, U ($^{233}U + ^{235}U \geq 20\%$); Pu, HEU and/or ^{233}U in scrap or other miscellaneous impure compounds.
c. Pu, HEU or ^{233}U in irradiated fuel.
d. U containing <20% ^{233}U and/or ^{235}U; Th.

Data Source: *IAEA Safeguards Glossary*, 2001 Ed., International Nuclear Verification Series No. 3, Vienna, Austria: IAEA, June 2002, sec. 3, paras. 13 and 20.

Table 2. IAEA's Estimated Conversion Time vs. Timeliness Detection Goal.

To take the most time-sensitive proliferation scenario, if a non-nuclear-weapon state has acquired at least one SQ of highly enriched uranium, uranium-233, or separated plutonium in metallic form, then the IAEA estimates that this state requires roughly seven-to-ten days to prepare its unirradiated, direct-use fissile material for insertion into a nuclear weapon.[24] Yet in terms of detecting a non-nuclear-weapon state's diversion of such material, the IAEA sets its *timeliness detection goal* as one month.[25]

Other plausible proliferation scenarios raise alarms because they illustrate just how easily non-nuclear-weapon states, through the possession of overt or covert nuclear fuel-making technologies, can clandestinely acquire weapons-ready nuclear material long before the IAEA is able to detect the acquisition. For example, reprocessing experts from Oak Ridge National Laboratory showed in an August 1977 technical brief how a non-nuclear-weapon state—which possesses irradiated direct-use materials, such as the sort of plutonium-laden spent nuclear fuel generated by light water reactors (LWRs)—could build, using simple industrial tools and a compact facility, a concealed "quick and dirty" reprocessing plant.[26] To take another example, former Nuclear Regulatory Commissioner Victor Gilinsky, MIT professor Marvin Miller, and former weapons-lab physicist Harmon Hubbard described in a 2004 report the relative ease and rapidity with which a state that possesses declared or clandestine centrifuge enrichment capability as well as nuclear fuel containing low enriched uranium (LEU), an "indirect-use" material used in LWRs, could enrich without detection this LEU to weapons-usable HEW.[27] "It is now generally appreciated that gas centrifuge plants for LEU can fairly easily be turned into plants for HEU," Gilinsky and company explained. "It is less appreciated that LEU at, say, 4 percent enrichment, is about 80 percent of the way to HEU. It takes comparatively little additional 'separative work' to upgrade LEU to HEU. It would be difficult for the IAEA to keep close enough track of all the LEU to stay ahead of any such conversion."[28]

Nuclear Facilities: Detecting Abrupt and Protracted Diversions.

When administering safeguards on nuclear material at facilities, the Agency has further translated its detection goals into what it now terms the *IAEA inspection goal,* defined as "[p]erformance targets specified for IAEA verification activities at a given facility as required to implement the facility safeguards approach."[29] To determine the IAEA inspection goal, the Agency uses the concept of a "material balance period"—that is, the amount of time between inventory accounts of declared nuclear materials at a given facility[30]—to clarify the two sorts of diversions-over-time that can occur at facilities: *abrupt diversions,* which occur when "the amount diverted is 1 SQ or more of nuclear material in a short time (i.e., within a period that is less than the material balance period)"; and *protracted diversions,* which occur when "the diversion of 1 SQ or more occurs gradually over a material balance period, with only small amounts removed at any one time."[31] The IAEA inspection goal at facilities thus consists of two corresponding components: the *timeliness component,* which "relates to the periodic activities that are necessary for the IAEA to be able to draw the conclusion that there has been no *abrupt diversion* of 1 SQ or more at a facility during a calendar year";[32] and the *quantity component,* which "relates to the scope of the inspection activities at a facility that are necessary for the IAEA to be able to draw the conclusion that there has been no [*protracted*] *diversion* of 1 SQ or more of nuclear material over a material balance period and that there has been no undeclared production or separation of direct use material at the facility over that period."[33]

However, when the Agency administers safeguards on nuclear material at so-called "bulk-handling" facilities—such as "plants for conversion, enrichment (or isotope separation), fuel fabrication and spent fuel reprocessing, and storage facilities for bulk material,"[34] it sometimes faces difficulties in meeting the IAEA inspection goal. In using materials accountancy to establish the timeliness and quantity components of the IAEA inspection goal at facilities, the Agency generally assumes a "detection probability" of 95 percent, a corresponding "false alarm probability" of 5 percent, and a measurement error of ± 1 percent. The *false alarm probability*, which the IAEA defines as "[t]he probability . . . that statistical analysis of accountancy verification data would indicate that an amount of nuclear material is missing when, in fact, no diversion has occurred,"[35] depends on both the estimated total amount of nuclear material going through the facility during an interval of time, and the threshold amount of the facility's nuclear material that the Agency must measure as missing during this time interval before it will begin suspecting a diversion.

The serious risks raised by abrupt diversion were outlined in the section discussing the gap between the conversion times of various nuclear materials and the respective IAEA's *timeliness detection goal* metrics for such materials. With respect to protracted diversions at facilities, the Agency faces even more serious difficulties in determining whether or not the "measured" missing nuclear material is explained by simply a measurement error or, since the quantity diverted from the facility at any one time over the material balance period need only be small compared to the absolute amount of material accounted for during the period, by an actual protracted diversion.

In a 1990 essay, MIT professor Marvin Miller offers an example in which a state operates a commercial-sized plutonium reprocessing plant through which 800 metric tons of spent nuclear fuel passes annually. To arrive at a *false alarm probability* of no more than five percent and a corresponding *detection probability* of 95 percent at such a plant over a 1-year material balance period, Miller calculates that the IAEA would have tolerate annually as much as 246 kilograms of "measured" missing plutonium—*an amount equivalent to over 30 significant quantities (or nuclear weapons-worth) of plutonium!*[36]

Yet, even if the IAEA should detect a sufficiently large discrepancy pointing potentially to protracted diversion at a nuclear fuel-making facility, resolution of this discrepancy would be far from timely. "If a large discrepancy is detected, the Agency will have to spend months working with the plant operator to figure out the technical reason for the discrepancy, prior to officially declaring the discrepancy an anomaly that needs to be resolved," observed Paul Leventhal in a 1994 essay. "The process of resolving an anomaly to the point of determining whether a suspected diversion should be reported to the IAEA Board of Governors could take months more, as could the process of the Board determining whether the matter needs to be referred back to the [IAEA] inspectors for further resolution or is of a magnitude to be referred to the UN Security Council."[37] With good reason, then, did Dr. Pierre Goldschmidt, the former IAEA Deputy Director General for Safeguards and Verification, concede after leaving the Agency that "there are still problems inherent in ensuring that, in 'bulk facilities,' even small amounts of nuclear material—a few kilograms among tons—are not diverted without timely warning."[38]

Moreover, as Union of Concerned Scientists' Edwin Lyman, one of many analysts today arguing for the IAEA to accept higher *false alarm probabilities*, recently noted, "The Agency's reluctance to pursue higher confidence levels for detection of diversion, at the expense of higher false alarm rates, would seem to be a lesser concern in the context of the heightened security levels that have become standard operating practice around the world since the 9/11 [September 11, 2001] attacks." He added:

> Today, most people are willing to tolerate a level of sensitivity for security screening at airports and critical facilities that would not have been acceptable in the past because of a common appreciation that the occasional false alarm is an appropriate price to pay to ensure that policy of as close to zero-tolerance as possible for the prevention of another 9/11-scale terrorist attack. Similarly, the standards for assurance that safeguards on plutonium used in the civil sector will be stringent enough to ensure an extremely high level of deterrence against diversion or theft should likewise be increased today, *yet it has not been* (emphasis added).

"On the contrary," Lyman lamented, "a growing appreciation of the inability of current measures to meet quantitative detection goals have led to a retreat from the notion that such goals should even be considered as standards for future achievement."[39] He explained:

> Although society may tolerate small leaks from a chemical plant to the environment if the hazards are limited, when the material in question can be used to build nuclear weapons, there is no acceptable level of

leakage into the hands of hostile states or terrorists. The consequences of a single nuclear weapon falling into the wrong hands would be so catastrophic that there must be a zero-tolerance policy for diversion.[40]

If the very standards which the IAEA has established for safeguarding nuclear fuel-making cannot be met, then claims that the entire nuclear fuel-cycle can be effectively safeguarded deserve to be not merely questioned, but also directly challenged.

Sufficiency of IAEA Resources.

The extent to which the IAEA actually possesses sufficient financial resources to perform its mission remains unclear. In turn, this uncertainty points to the larger issue of whether the Agency can effectively safeguard nuclear activities, and thus verify the fulfillment of NPT obligations by signatories.[41]

In September 2006, Henry Sokolski, executive director of the Nonproliferation Policy Education Center, warned of the growing gap between IAEA resources and safeguarding responsibilities when he testified before the U.S. House of Representatives' Subcommittee on National Security, Emerging Threats, and International Relations. Table 3, which the author assisted Sokolski in preparing, gives the figures on the IAEA's safeguards budget obligation in constant dollars, and amounts of unirradiated direct-use nuclear materials for the years 1984 and 2004.[42]

	As of 1984	As of 2004
IAEA Safeguards Budget Obligation (In Constant Fiscal Year 2004 U.S. Dollars)	$45.7 million	$104.9 million
Separated Plutonium (Pu) Outside Reactor Cores	7.7 tonnes	89.0 tonnes
High Enriched Uranium (HEU)	11.8 tonnes	32.0 tonnes
Total IAEA Safeguarded Weapons-Usable Nuclear Materials	19.5 tonnes	121.0 tonnes
Data Sources: For data on the IAEA's safeguards budget obligation in current—not constant—U.S. dollars, see *The Agency's Accounts for 1984*, GC(XXIX)/749, p. 26; and *The Agency's Accounts for 2004*, GC(49)/7, p. 47. For data on the amount of nuclear material safeguarded by the IAEA, *see Annual Report for 1984*, GC(XXIX)/748, p. 63; and *Annual Report for 2004*, GC(49)/5, Annex, Table A19, Vienna, Austria: IAEA, July 1985.		

Table 3. IAEA Safeguards Budget, and Safeguarded Weapons-Usable Nuclear Materials in Non-Nuclear-Weapon Signatories of the NPT.

Over a 20-year period, the IAEA's safeguards and verification budget only roughly doubled in constant dollars, while civilian stockpiles of plutonium and highly-enriched uranium in non-nuclear-weapon States—unirradiated weapons-ready nuclear materials for which the Agency must account—increased by a factor of six.

Figure 1 graphically illustrates the IAEA's safeguards budget obligation in constant fiscal year 2000 U.S. dollars from 1970, the year when the NPT entered into force, to 2005. As the graph shows, after 1995 the IAEA safeguards budget obligation did not just experience zero real growth, but rather contracted significantly, and began only within

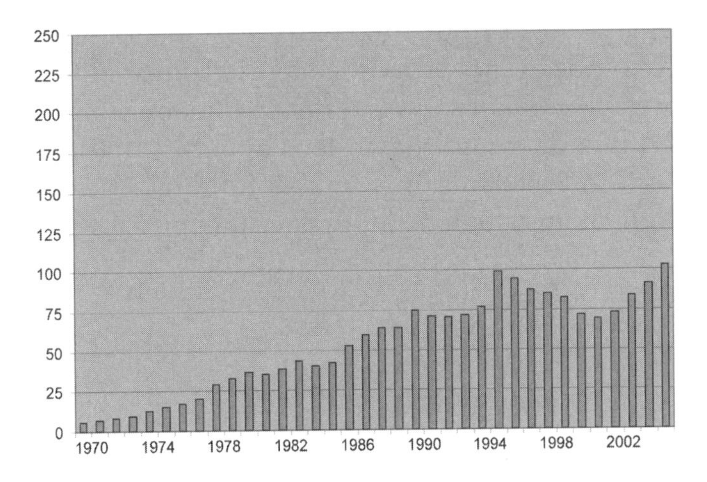

Data Sources: For data on the IAEA's safeguards budget obligation in current—not constant—U.S. dollars, see *The Agency's Accounts for 1970*, GC(XV)/459, Vienna, Austria: IAEA, July 1971, p. 12; *The Agency's Accounts for 1971*, GC(XVI)/484, Vienna, Austria: IAEA, July 1972, p. 16; *The Agency's Accounts for 1972*, GC(XVII)/504, Vienna, Austria: IAEA, August 1973, p. 16; *The Agency's Accounts for 1973*, GC(XVIII)/527, Vienna, Austria: IAEA, August 1974, p. 16; *The Agency's Accounts for 1974*, GC(XIX)/549, Vienna, Austria: IAEA, August 1975, p. 13; *The Agency's Accounts for 1975*, Draft, GOV/1781, Vienna, Austria: IAEA, April 1976, p. 11; *The Agency's Accounts for 1976*, GC(XXI)/581, Vienna, Austria: IAEA, June 1977, p. 13; *The Agency's Accounts for 1977*, GC(XXII)/598, Vienna, Austria: IAEA, July 1978, p. 13; *The Agency's Accounts for 1978*, GC(XXIII)/611, Vienna, Austria: IAEA, August 1979, p. 16; *The Agency's Accounts for 1979*, GC(XXIV)/629, Vienna, Austria: IAEA, July 1980, p. 18; *The Agency's Accounts for 1980*, GC(XXV)/645, Vienna, Austria: IAEA, July 1981, p. 18; *The Agency's Accounts for 1981*, GC(XXVI)/665, Vienna, Austria: IAEA, July 1982, p. 20; *The Agency's Accounts for 1982*, GC(XXVII)/685, Vienna, Austria: IAEA, August 1983, p. 20; *The Agency's Accounts for 1983*, GC(XXVIII)/714, Vienna, Austria: IAEA, August 1984, p. 20; *The Agency's Accounts for 1984*, GC(XXIX)/749, Vienna, Austria: IAEA, August 1985, p. 26; *The Agency's Accounts for 1985*, GC(XXX)/776, Vienna, Austria: IAEA, August 1986, p. 30; *The Agency's Accounts for 1986*, GC(XXXI)/801, Vienna, Austria: IAEA, August 1987, p. 46; *The Agency's Accounts for 1987*, GC(XXXII)/836, Vienna, Austria: IAEA, August 1988, p. 46; *The Agency's Accounts for 1988*, GC(XXXIII)/874, Vienna, Austria: IAEA, August 1989, p. 42; *The Agency's Accounts for 1989*, GC(XXXIV)/916, Vienna, Austria: IAEA, July 1990, p. 44; *The Agency's Accounts for 1990*, GC(XXXV)/954, Vienna, Austria: IAEA, August 1991, p. 52; *The Agency's Accounts for 1991*, GC(XXXVI)/1005, Vienna, Austria: IAEA, August 1992, p. 68; *The Agency's Accounts for 1992*, GC(XXXVII)/1061, Vienna, Austria: IAEA, August 1993, p. 24; *The Agency's Accounts for 1993*, GC(XXXVIII)/4, Vienna, Austria: IAEA, August 1994, p. 18; *The Agency's Accounts for 1994*, GC(39)/5, Vienna, Austria: IAEA, August 1995, p. 22; *The Agency's Accounts for 1995*, GC(40)/9, Vienna, Austria: IAEA, August 1996, p. 36; *The Agency's Accounts for 1996*, GC(41)/9, Vienna, Austria: IAEA, August 1997, p. 34; *The Agency's Accounts for 1997*, GC(42)/6, Vienna, Austria: IAEA, August 1998, p. 50; *The Agency's Accounts for 1998*, GC(43)/5, Vienna, Austria: IAEA, August 1999, p. 44; *The Agency's Accounts for 1999*, GC(44)/5, Vienna, Austria: IAEA, July 2000, p. 50; *The Agency's Accounts for 2000*, GC(45)/7, Vienna, Austria: IAEA, August 2001, p. 46; *The Agency's Accounts for 2001*, GC(46)/6, Vienna, Austria: IAEA, July 2002, p. 32; *The Agency's Accounts for 2002*, GC(47)/4, Vienna, Austria: IAEA, August 2003, p. 44; *The Agency's Accounts for 2003*, GC(48)/9, Vienna, Austria: IAEA, August 2004, p. 39; *The Agency's Accounts for 2004*, GC(49)/7, Vienna, Austria: IAEA, August 2005, p. 47; and *The Agency's Accounts for 2005*, GC(50)/8, Vienna, Austria: IAEA, July 2006, p. 54.

To convert the current U.S. dollars (USD) into crudely estimated constant FY2000 USD, the author inferred deflators from data of the U.S. Department of Defense's Office of the Comptroller and the Congressional Budget Office.

Figure 1. IAEA Safeguard Budget Obligations, 1970-2005. (Estimated Constant FY2000 $USD Millions)

recent years to return to mid-1990s spending levels. Such trends in IAEA funding have led nonproliferation experts like Sokolski to call for drastic revisions to the IAEA's budget and the system by which the Agency assess governments for annual funding. "If we are serious about safeguarding against the spread of nuclear weapons and preventing nuclear theft or terrorism," Sokolski told the House subcommittee, "these trends [of under-funding the IAEA] must change." In fact, in October 2006 IAEA Director General Mohamed ElBaradei himself argued emphatically for more Agency resources:

> Our [safeguards] budget is only 130 million dollars. That's the budget with which we're supposed to verify the nuclear activities of the entire world . . . . Our budget, as I have said before, is comparable with the budget of the police department in Vienna. So we don't have the required resources in many ways to be independent, to buy our own satellite monitoring imagery, or crucial instrumentation for our inspections. We still do not have our laboratories here in Vienna equipped for state-of the-art analysis of environmental samples.[43]

At a minimum, ElBaradei's argument suggests the need for more transparent discussion of the extent to which the IAEA, given its limitations in financial and other resources, is capable of administering effective safeguards worldwide.

In sum, the analysis of this section suggests that the IAEA, given its own safeguarding goals, remains unable to safeguard effectively a broad range of sensitive nuclear materials and activities. In particular, the Agency cannot provide—even in principle!—timely warning of a non-nuclear-weapon state's abrupt diversion of the most weapons-ready nuclear materials (i.e., highly enriched uranium and plutonium) from civilian applications to nuclear weapons or purposes unknown. Moreover, under current accounting methods dictating a false alarm rate of at most five percent, the IAEA must tolerate many significant quantities of unaccounted nuclear material at bulk-handling facilities before even suspecting a protracted diversion. Finally, the Agency appears to lack adequate financial resources to carry out many of its safeguarding activities effectively.

Taken together, these findings raise important questions:

- Does the NPT's Article IV affirm the right of signatories to nuclear materials and activities that the IAEA cannot effectively safeguard?
- More broadly, does Article IV affirm the right of signatories to peaceful nuclear energy without any qualifications whatsoever?

These questions go to the heart of debates over the precise meaning of the "research, production, and use of nuclear energy for peaceful purposes without discrimination and in conformity with Articles I and II [of the NPT]," and the "fullest possible exchange of equipment, materials, and scientific and technological information for the peaceful uses of nuclear energy," to which treaty signatories have a right under Article IV. Answers to these questions require a sustained analysis and interpretation of the NPT itself.

CLARIFYING THE SCOPE AND LIMITS OF THE NPT'S ARTICLE IV

Treaties demand careful interpretation, and the Nuclear Nonproliferation Treaty is no exception. Careful interpretation is demanded because treaties sometimes contain *ambiguous* language and, as Fred C. Iklé noted in his important 1964 study on negotiation, treaty language that lacks specificity can lead parties to "have an honest misunderstanding about implications that the agreement fails to spell out"; or one party, "while knowing what its opponent expected of the bargain," to "pretend that it had a different understanding of it (i.e., the ambiguities are exploited to cover up a deliberate violation.)"[44] Careful interpretation is also demanded because treaties may sometimes contain *equivocal* language. According to Iklé, such equivocality occurs when:

> the parties to the agreement *know* that the ambiguous terms mean different things to each of them . . . . Equivocal language is used to *cover up* disagreement on issues which must be included for some reason in a larger settlement or which must be dealt with as if there was agreement. An equivocal agreement is similar to a partial agreement that leaves certain undecided issues for future negotiation, with the difference that the equivocal terms serve to cover up differences rather than mark them for future resolution (emphasis in the original).[45]

When governments interpret treaties to deal with issues of ambiguity and equivocality, they generally adhere to a set of internationally-accepted principles that Section Three of the *Vienna Convention on the Law of Treaties* (VCLT) seeks to codify.[46] Article 31 of the VCLT identifies the primary means of interpretation as the close reading of a treaty "in good faith in accordance with the ordinary meaning to be given to the terms of the treaty in their context and in the light of [the treaty's] object and purpose."[47] In addition, the VCLT's Article 32 endorses the use of "supplementary means of interpretation," such as a treaty's negotiation history and other *travaux préparatoires* (preparatory materials), in order to confirm an Article 31-derived interpretation, or to determine a treaty's meaning when such an interpretation "leaves the meaning ambiguous or obscure" or "leads to a result which is manifestly absurd or unreasonable."[48] To the extent that governments advocating a per se right or unqualified per se right interpretation of the NPT's Article IV have arrived at this reading using the means of interpretation codified by the VCLT, the VCLT's Article 32 provides a warrant for recourse to the NPT's negotiation history, at the very least, to confirm whether or not this history supports this reading.

Negotiating and Concluding the NPT.

The multilateral negotiations that led eventually to the NPT's conclusion took place during the mid-to-late 1960s in several contexts.[49] Among the most important of these was the Eighteen Nation Disarmament Committee (ENDC).[50] Formed in late 1961, the ENDC consisted of five states from the West: Britain, Canada, France, Italy, and the United States; five states from the Soviet bloc: Bulgaria, Czechoslovakia, Poland, Romania, and the Union of Soviet Socialist Republics (USSR); and eight nonaligned states: Brazil, Burma, Ethiopia, India, Mexico, Nigeria, Sweden, and the United Arab Republic.[51] (France, however, declined to participate in the ENDC.)

When the ENDC began meeting in Geneva, Switzerland, in March 1962, it initially set out to negotiate and conclude an agreement on "general and complete disarmament under effective international control."[52] Over the next few years, though, negotiations stalled as American and Soviet delegates continually found themselves at loggerheads. But after the People's Republic of China's surprise detonation of a nuclear explosive device in October 1964, ENDC delegates changed the focus of their negotiations to concluding a nuclear nonproliferation treaty.[53]

Prior to August 24, 1967, no draft nuclear nonproliferation treaty submitted to the ENDC contained any language whatsoever viewing peaceful uses of nuclear energy through the prism of "legal rights." The idea for treaty language affirming the "rights" of signatories to peaceful nuclear energy apparently came from the *Treaty for the Prohibition of Nuclear Weapons in Latin America and the Caribbean* (Treaty of Tlatelolco), Article 17 of which states "Nothing in the provisions of this Treaty shall prejudice the rights of the Contracting Parties, in conformity with this Treaty, to use nuclear energy for peaceful purposes, in particular for their economic development and social progress."[54] The negotiations for the Treaty of Tlatelolco took place in the mid-to-late 1960s within the context of the Preparatory Commission for the Denuclearization of Latin America (known also by its Spanish acronym, COPREDAL). According to a confidential telegram from the U.S. Embassy in Mexico to the Department of State, the Peruvian delegation first proposed to COPREDAL the idea of including an article on nuclear energy for peaceful purposes.[55]

Soon after the conclusion of the Treaty of Tlatelolco's negotiations in mid-February 1967, Latin American delegations to the ENDC began proposing that the draft nuclear nonproliferation treaty include language similar to Tlatelolco's Article 17. The following month, American and Soviet negotiators began privately discussing possible language to deal with the issue of nuclear rights. As a mid-April 1967 memorandum suggests, originally the United States proposed to the USSR that language dealing with nuclear rights appear in the preamble, and later proposed placing such language at the end of Article III, the provision on international control and IAEA safeguards. Throughout these private bilateral consultations, though, U.S.-origin draft language did not propose to treat nuclear-rights language in terms of "inalienable rights."[56] According to an internal, now-declassified negotiation history from the U.S. Arms Control and Disarmament Agency, it was the Soviets who first proposed using the phrase "inalienable right" in Article IV:

> While we originally preferred to leave the question of specific treaty language to nonaligned initiative during later negotiations, we agreed in May [1967] to a brief Soviet draft article:
>
> Article IV
>
> Nothing in the Treaty shall be interpreted as affecting the inalienable right of all the Parties to develop research, production and use of nuclear energy for peaceful purposes without discrimination and in conformity with Articles I and II of this treaty, as well as the right of the Parties to participate in the fullest possible exchange of information for, and to contribute alone or in co-operation with other States to, the further development of the applications of nuclear energy for peaceful purposes.[57]

In late August 1967, the American and Soviet delegations, after months of consultations with each other and with officials from other governments, tabled in the ENDC identical nonproliferation treaty drafts. These drafts contained this first, and a much shorter, version of Article IV that is quoted above.[58]

Over the next year, ENDC delegates struggled to refine (among other things) the language of the NPT's Article IV. At each turn, though, they collectively rejected several proposals to insert language into the treaty that would have expressly and explicitly recognized the per se right of signatories to the supply of enrichment, reprocessing and other sensitive nuclear fuel-making technologies—and, in certain cases, to the acquisition of so-called nuclear explosive devices for civilian purposes. Several proposals stand out:

- In September 1967, Mexican delegate Jorge Castañeda proposed to the ENDC that Article IV includes a second paragraph establishing "the duty" (or express legal obligation) of "[t]hose parties that are in a position to do so . . . to contribute, according to their ability, alone or in cooperation with other States or international organizations, to the further development of the production, industries, and other applications of nuclear energy for peaceful purposes, especially in the territories of non-nuclear-weapon States."[59]
- In mid-October 1967, the Romanian delegation to the ENDC submitted a working paper suggesting the inclusion of language in the preamble recognizing the right of signatories to nuclear energy for peaceful purposes as an "absolute right," which is to say, an unqualified right.[60]
- In late October 1967, the Brazilian delegation to the ENDC offered its own working paper proposing that Article IV expressly recognize the "inalienable right" of signatories to develop not only "nuclear energy for peaceful purposes," but all nuclear technologies (presumably including nuclear fuel-making) up to "nuclear explosive devices for civil uses."[61]
- In early November 1967, the Nigerian delegation (in what appears to be an elaboration and

extension of the obligatory "duty" language found in Mexico's September 1967 working paper) proposed that Article IV add several paragraphs that would legally oblige transfers of nuclear material and technology.[62]

- In early February 1968, the Spanish government, which was not a member of the ENDC, submitted a memorandum to the committee calling for Article IV's second paragraph to refer expressly to nuclear fuel-making technologies:

> The measures in the new draft concerning the right to participate as fully as possible in scientific and technical information for the peaceful uses of atomic energy are sound, and can have important effects on the development of non-nuclear countries. Nevertheless, the Spanish Government takes the view that this information should refer *specifically to the entire technology of reactor and fuels* [emphasis added].[63]

- In mid-February 1968, the Brazilian delegation once again proposed that Article IV's first paragraph affirm the right of all signatories to develop not only "nuclear energy for peaceful purposes," but also all nuclear technology up to so-called peaceful nuclear explosive devices.[64]
- And in late February 1968, the Italian delegation proposed that Article IV's second paragraph be revised to contain instead the following negative declaration:

> Nothing in this Treaty shall be interpreted as affecting the inalienable right of all the Parties to *the supply of source and special fissionable materials or equipment for the use of source and special fissionable materials for peaceful purposes.*[65]

All of these proposals were rejected by the ENDC. Indeed, the final text of the NPT contained no language explicitly referring to enrichment, reprocessing, and other nuclear fuel-making activities, or so-called nuclear explosive devices for peaceful purposes—let alone expressly recognizing the per se right of signatories to nuclear fuel-making.[66]

That said, the NPT's Article III contains language that, at the very least, appears to contemplate that both nuclear-weapon and non-nuclear-weapon signatories might produce, access, and use even the most weapons-ready nuclear materials in civilian applications of nuclear technology. Article III's first paragraph states:

> Procedures for the safeguards required by this Article shall be followed with respect to source or special fissionable material whether it is being produced, processed or used in any principal nuclear facility or is outside any such facility. The safeguards required by this Article shall be applied on all source or special fissionable material in all peaceful nuclear activities within the territory of such State, under its jurisdiction, or carried out under its control anywhere.[67]

Paragraph two adds:

> Each State Party to the Treaty undertakes not to provide: (a) source or special fissionable material, or (b) equipment or material especially designed or prepared for the processing, use or production of special fissionable material, to any non-nuclear-weapon State for peaceful purposes, unless the source or special fissionable material shall be subject to the safeguards required by this Article.[68]

Yet, though these provisions describe the scope of responsibility for IAEA safeguards, they do not explicitly address the range of nuclear activities that are prohibited or permitted. Rather, the key to harmonizing these provisions with the NPT's larger prohibitions against proliferation in Articles I and II lies in Article III's third paragraph, which states: "The safeguards required by this Article shall be implemented in a manner designed to comply with Article IV of this Treaty . . ."[69] If Article IV is read broadly and permissively to permit any nuclear activity short of inserting fissile material into a nuclear explosive, then the IAEA safeguards required by Article III will play, at best, a formalistic role, and the importance of the effectiveness (or lack of effectiveness) of IAEA safeguards will be of little consequence. In contrast, if Article IV—as well as Articles I and II, the NPT provisions to which Article IV itself *shall conform*—are read carefully and less permissively, then the actual effectiveness of IAEA safeguards will play a crucial role in determining whether or not certain types of nuclear materials, technologies and activities should enjoy protection under Article IV.

The Three Qualifications of Article IV's "Inalienable Right."

To be sure, Article IV of the NPT recognizes the "inalienable right" of signatories to peaceful nuclear energy. However, it also explicitly imposes two qualifications on the "nuclear energy for peaceful purposes" to which NPT signatories have an "inalienable right." Signatories shall develop "research, production, and use" of peaceful nuclear energy (1) "without discrimination," and (2) "in conformity with articles I and II of this Treaty."[70] Moreover, when the NPT's Article III defines the purpose of comprehensive safeguards by the IAEA as the "*verification of the fulfillment of* [*signatory*] *obligations assumed under this Treaty* with a view to preventing the diversion of nuclear energy from peaceful purposes to nuclear weapons and other nuclear explosive devices,"[71] it effectively establishes (3) "conformity with Article III" as a third qualification. These three qualifications, when understood in relation to the treaty's preamble and main text, not only narrow the scope of "nuclear energy for peaceful purposes" to which signatories have an "inalienable right," but also establish criteria that signatories must meet in order to exercise this right.

To begin with, paragraph seven of the NPT's preamble lays out the principle that addresses the special meaning of Article IV's first qualification, "without discrimination," within the context of the treaty.[72] That paragraph affirms:

> the principle that *the benefits* of peaceful applications of nuclear technology, including any technological by-products which may be derived by nuclear-weapon States from the development of nuclear explosive devices, should be available for peaceful purposes to all Parties of the Treaty, whether nuclear-weapon or non-nuclear weapon States (emphasis added).[73]

To be clear, neither this principle (which hereinafter I refer to as the "benefits-without-discrimination" principle), nor any other part of the NPT, ever expressly requires that any specific *nuclear technology*, or any specific *peaceful application* of nuclear technology, be made available to all signatories, but rather that only that *the benefits* of a given nuclear technology's peaceful application be made available somehow. In essence, this principle recognizes that some nuclear technologies and some peaceful applications of nuclear technology—to take an extreme example, so-called "nuclear explosions for peaceful purposes" in civilian mining, excavation, or canal-digging operations—

may be too uneconomical, too proliferative, and too unsafeguardable to permit non-nuclear-weapon states to acquire and use them. Thus, when Article IV's first qualification applies this principle to peaceful nuclear energy, it appears to permit, in principle, the denial of a given nuclear technology or a given nuclear technology's peaceful application to a signatory as long as the benefits of the denied nuclear technology's peaceful application are made available somehow.

Article IV's second qualification requires that the development of "research, production, and use" of peaceful nuclear energy be "in conformity with articles I and II" of the NPT. These two articles articulate the NPT's main prohibitions against the direct and indirect proliferation of nuclear weapons by treaty signatories. Article I prohibits nuclear-weapon signatories from giving nuclear weapons and other nuclear explosive devices, or control over such devices, to "any recipient whatsoever," and also forbids them from "assist[ing], encourage[ing], or induc[ing]" any non-nuclear-weapon state "to manufacture or otherwise acquire" nuclear explosive devices.[74] Article II correspondingly prohibits non-nuclear-weapon signatories from receiving nuclear explosive devices, or control over such devices, and also forbids them from building or acquiring in any way nuclear explosive devices, and from receiving or seeking "any assistance in the manufacture" of such devices.[75] Article IV's second qualification therefore effectively narrows the scope of "nuclear energy for peaceful purposes" to which signatories have an "inalienable right" under Article IV, for peaceful nuclear energy "in conformity with articles I and II" excludes *not only* nuclear explosive technology for peaceful or nonpeaceful purposes, *but also* other nuclear technology and assistance that could "assist, encourage, or induce" non-nuclear-weapon states "to manufacture or otherwise acquire" nuclear explosive technology.[76]

Furthermore, the NPT's Article III requires each non-nuclear-weapon signatory to conclude a comprehensive safeguard agreement with the IAEA "for the exclusive purpose of *verification of the fulfillment of its obligations assumed under this Treaty* with a view to preventing the diversion of nuclear energy from peaceful purposes to nuclear weapons and other nuclear explosive devices" (emphasis added).[77] By requiring non-nuclear-weapon signatories to submit to full-scope IAEA safeguards in order to verify the fulfillment of their obligations under Articles I and II, as well as other parts of the NPT, Article III effectively establishes a third legally-binding qualification on the "nuclear energy for peaceful purposes" to which signatories have an "inalienable right" under Article IV. That is, to develop "research, production, and use" of peaceful nuclear energy "in conformity with articles I and II" necessarily implies full "conformity with article III."[78] Thus, Article IV's third qualification appears to recognize the "inalienable right" of a signatory to peaceful nuclear energy only when the signatory's nuclear activities are *effectively* safeguardable by the IAEA, and the signatory complies fully with its obligations under Article III of the NPT and related IAEA comprehensive safeguards agreements.[79]

Article IV's Three Qualifications and Nuclear Explosions for Peaceful Purposes.

With respect to "nuclear explosions for peaceful purposes," the majority of the NPT negotiators understood that, at the time of their negotiations and for the foreseeable future, nuclear explosive technology in civilian projects not only lacked clear and immediate economic benefits, especially when compared to non-nuclear alternatives; but also possessed an unacceptable risk of nuclear proliferation since such technology *could not be effectively safeguarded by the IAEA*. Hence, the final text of the NPT denies non-nuclear-weapon signatories access both to nuclear explosive technology and its peaceful applications.

In conformity with the preamble's "benefits-without-discrimination" principle, though, the NPT's Article V outlines the framework by which non-nuclear-weapon signatories could avail themselves of "the potential benefits" of nuclear explosive technology's peaceful application, if such economic benefits should ever materialize. The relevant part of Article V reads:

> Each Party to the Treaty undertakes to take appropriate measures to ensure that, in accordance with this Treaty, under appropriate international observation and through appropriate international procedures, *potential benefits* from any peaceful applications of nuclear explosions will be made available to non-nuclear-weapon States Party to the Treaty on a *non-discriminatory basis* and that the charge to such Parties for the explosive devices used will be as low as possible and exclude any charge for research and development. Non-nuclear-weapon States Party to the Treaty shall be able to obtain such benefits, pursuant to a special international agreement or agreements, through an appropriate international body with adequate representation of non-nuclear-weapon States . . . (emphasis added).[80]

As the NPT's negotiation history reveals, many of the non-nuclear-weapon states represented at the ENDC did not view either the denial of nuclear explosive technology and its peaceful applications, or Article

V's framework for providing the "potential benefits" of the denied nuclear explosive technology's peaceful applications, as discriminatory per se.[81] For example, in late January 1968 Polish delegate Mieczyslaw Blusztajn remarked to the ENDC:

> I should like once again to stress that *the right of all countries to conduct peaceful nuclear explosions is not at stake.* The only matter to be settled is the procedure and the conditions to be observed so that countries which forgo the manufacture of nuclear devices shall not be deprived of *the benefits* that may be derived from the use of nuclear explosives (emphasis added).[82]

Bulgarian delegate Kroum Christov echoed the Polish delegate's sentiments:

> [I]t seems to us quite clearly impossible to admit and to include in the non-proliferation treaty the right to manufacture nuclear devices and to carry out nuclear explosions. *There is no question in this case of denying a right; nor should the prohibition of all activity of this nature be regarded as an infraction of that right.* Account is taken of a state of facts which, for reasons which cannot be refuted and which have been explained here at length, renders the manufacture of nuclear devices incompatible with a non-proliferation treaty (emphasis added).[83]

In retrospect, the efforts of NPT negotiators to limit the spread of nuclear explosive technology for peaceful purposes proved to be well-founded. Indeed, the "potential benefits" of so-called peaceful nuclear explosives (PNEs) never materialized as non-nuclear explosive alternatives for mining, excavation, and canal-digging operations emerged as safer and more economical choices.[84] In fact, in May 1995 the quadrennial NPT review conference made the following conclusions about PNEs:

> The Conference records that *the potential benefits* of the peaceful applications of nuclear explosions envisaged in article V of the Treaty *have not materialized*. In this context, the Conference notes that the potential benefits of the peaceful applications of nuclear explosions have not been demonstrated and that serious concerns have been expressed as to the environmental consequences that could result from the release of radioactivity from such applications and on the risk of possible proliferation of nuclear weapons. Furthermore, no requests for services related to the peaceful applications of nuclear explosions have been received by IAEA since the Treaty entered into force. The Conference further notes that no State party has an active programme for the peaceful application of nuclear explosions (emphasis added).[85]

Moreover, though the *Comprehensive Nuclear-Test-Ban Treaty* has not entered into force, it has nonetheless helped to support an international norm against the use of nuclear explosions, whether for nonpeaceful or allegedly peaceful purposes.[86]

By prohibiting non-nuclear-weapon states from developing, accessing, and using so-called peaceful nuclear explosive devices, the NPT reinforces the importance of the following principle: When the IAEA cannot effectively safeguard the nuclear material involved in an allegedly-peaceful application of nuclear technology, then the NPT does not protect the right of states to develop, access or use that allegedly-peaceful application of nuclear technology.

Article IV's Three Qualifications and Nuclear Energy for Peaceful Purposes.

In conformity with Article IV's three qualifications, then, both (a) the "benefits-without-discrimination" principle of the NPT's preamble, and (b) the framework by which Article V allows non-nuclear-weapon signatories to avail themselves of the "potential benefits" of nuclear explosive technology's peaceful applications without providing them actual access to the technology or its peaceful application, can be applied to enrichment, reprocessing, and other sensitive nuclear fuel-making activities. It is both plausible and consistent for governments to interpret Article IV as affirming the "inalienable right" of nuclear signatories to develop "research, production, and use" of nuclear fuel making only to the extent that such nuclear fuel-making activities: (1) are economically beneficial in accordance with the treaty's preamble (Article IV's first qualification); (2) possess a low risk of proliferation in accordance with Articles I and II (Article IV's second qualification); and (3) are *effectively* safeguardable and undertaken in full compliance with NPT and IAEA safeguard obligations in accordance with Article III (Article IV's third qualification).[87] Moreover, it is both plausible and consistent with the treaty to deny signatories from developing, acquiring, and using nuclear fuel-making technologies (especially those which are related to nuclear materials that the IAEA cannot effectively safeguard) that can assist them in manufacturing nuclear weapons under some circumstances—at the very least, when they fail to comply with their obligations under the NPT's Article III and related IAEA safeguards agreements—as long as the benefits of peaceful applications of such nuclear fuel-making technologies are made available to them.[88]

As the NPT's negotiation history reveals, ENDC delegations from both nuclear-weapon states and non-nuclear-weapon states viewed nuclear fuel-making in a manner similar to nuclear explosives for peaceful purposes: that is, as potentially aiding and even constituting the manufacture of nuclear weapons. For example, in September 1962 British delegate Sir Michael Wright told the ENDC:

> The thing which is unique to a nuclear weapon is its warhead. And what is there in a nuclear warhead that is found in no other weapons? . . . It is the fissile material in the warhead; that is to say, the plutonium and uranium-235, the two fissile materials now most commonly used in nuclear weapons.
>
> *If we are to deal effectively with nuclear weapons we must concentrate on the fissile material which every nuclear weapon has and which no other weapon has* [emphasis added].[89]

To take another example, in February 1966 Swedish delegate Alva Myrdal argued before the ENDC:

> We could, of course, all agree that it is important to block the road to nuclear-weapon development as early as possible. But we must be aware that what we are facing is a long ladder with many rungs, and the practical question is: on which of these is it reasonable and feasible to introduce the international blocking? . . . *To prohibit just the final act of "manufacture" would seem to come late in these long chains of decisions* [emphasis added]. . . Could a middle link be found on which the prohibitory regulation should most definitely be focused?[90]

A month later, during a speech to the ENDC, Burmese delegate U. Maung Maung Gyi answered Myrdal's question:

> *An undertaking on the part of the non-nuclear weapon Powers not to manufacture nuclear weapons would in effect mean forgoing the production of fissionable material* [emphasis added] . . . and such production is the first essential step for the manufacture of these weapons and constitutes an important dividing line between restraint from and pursuit of the nuclear path.[91]

Proponents of the per se right or unqualified per se right reading of Article IV might counter the above reading by claiming that Article IV's second paragraph necessarily obliges signatories to transfer any and all nuclear technology, materials, and assistance—including nuclear fuel making—in an unqualified and unfettered manner. The relevant part of that paragraph states: "All the Parties of the Treaty undertake to facilitate, and have the right to participate in, the fullest possible exchange of equipment, materials, and scientific and technological for the peaceful uses of nuclear energy."[92] It is important to note, though, that this paragraph is carefully worded to call not for "the fullest exchange," but rather for only "the fullest *possible* exchange," and thus actually encourages NPT signatories to exchange nuclear technology, materials, and know-how with great care, caution, and restraint.[93] In May 2005, during a speech to the quadrennial NPT review conference, Christopher Ford (at the time the Principal Deputy Assistant Secretary of State for Verification, Compliance and Implementation) elaborated this point:

> The use of the term "fullest possible" is an acknowledgement that cooperation may be limited. Parties are not compelled by Article IV to engage in nuclear cooperation with any given state—or to provide any particular form of nuclear assistance to any other state. The NPT does not require any specific sharing of nuclear technology between particular States Party, nor does it oblige technology-possessors to share any specific materials or technology with non-possessors.[94]

"[T]o conform both to the overall objective of the NPT—strengthening security by halting nuclear proliferation and to any Article I and III obligations," Ford added, "supplier states must consider whether certain types of assistance, or assistance to certain countries, are consistent with the nonproliferation purposes and obligations of the NPT, other international obligations, and their own national requirements." NPT signatories, Ford concluded, "should withhold assistance if they believe that a specific form of cooperation would encourage or facilitate proliferation, or if they believe that a state is pursuing a nuclear weapons program in violation of Article II, is not in full compliance with its safeguards obligations, or is in violation of Article I."[95] Moreover, by establishing no per se obligation or duty of nuclear exporters to give any specific nuclear technology, material, or assistance, Article IV's second paragraph suggests that nuclear importers, at the same time, have no reciprocal per se right to receive or otherwise acquire any specific nuclear technology, material or assistance.[96]

In sum, the analysis of this section suggests that the NPT does not affirm the "inalienable right" of signatories to nuclear materials and activities that the IAEA cannot effectively safeguard. In fact, this explains why the treaty prohibits non-nuclear-weapon signatories from developing, accessing, or using so-called "nuclear explosions for peaceful purposes," the most military-relevant of civilian applications of nuclear technology. Instead, the NPT appears to establish three qualifications on Article IV which condition the extent to which signatories have an "inalienable right" to develop and use peaceful nuclear energy—key qualifications being a signatory's full compliance with its obligations under the NPT and IAEA comprehensive safeguards agreements and the actual ability of the IAEA to administer *effective* safeguards on nuclear materials in a given civilian application of nuclear technology. But while the NPT may be understood as prohibiting non-nuclear-weapon signatories from unsafeguardable nuclear materials and activities, the treaty also provides for mechanisms by which nuclear-weapon signatories can provide, individually or through multilateral frameworks, *the benefits* of proscribed, unsafeguardable peaceful applications of nuclear technology in a nondiscriminatory manner to non-nuclear-weapon signatories in full compliance.

MOVING AWAY FROM A CROWD OF VIRTUAL NUCLEAR-WEAPON STATES

On the morning of October 9, 2006, the Democratic People's Republic of Korea exploded a nuclear weapon.[97] Having long subscribed to the unqualified per se right reading of the Nuclear Nonproliferation Treaty, North Korea became the first ever non-nuclear-weapon state to use the treaty as cover for the overt and covert production of weapons-usable fissile material, and then to quit the treaty, and later build and detonate a nuclear explosive device.[98]

If governments continue to interpret the NPT as recognizing the per se right or, worse, the unqualified per se right of signatories to enrichment, reprocessing, and other sensitive nuclear activities, then this will all but guarantee the emergence of more nuclear fuel-making states—of what IAEA Director General Mohammed ElBaradei now chillingly describes as *virtual nuclear-weapon states*. The world will move towards a nuclear-armed crowd.

As we have seen, though, the NPT need not be read this way. Governments can—and should—interpret the treaty in a pragmatic and sustainable way that rejects not only claims of an unqualified per se right, but also of a per se right, of signatories to nuclear fuel making.

The Role of IAEA Candor.

A necessary condition for more pragmatic and sustainable readings of the NPT, however, will be IAEA candor with respect to what it can and cannot effectively safeguard. In the past, the Agency avoided discussion of this issue, but such avoidance served only to promote widely the mistaken belief that IAEA safeguards are always effective, even when applied to uranium enrichment, plutonium reprocessing, and other sensitive nuclear activities. In turn, this mistaken belief lent support to interpretations of the NPT's Article IV recognizing the per se right—and, in Iran's case, even the unqualified per se right—of NPT signatories to nuclear fuel-making.

The IAEA is certainly capable of candor. When the Persian Gulf War's aftermath exposed the extent to which the Agency could not verify the completeness of a state's declaration, the IAEA moved to clarify its legal inspection authority[99] and improve its technical capabilities.[100] To meet the dangers posed by the emergence of ever more *virtual nuclear-weapon states*, though, the Agency will have to do much more. In particular, the IAEA—given its inability at times to meet, in practice, key safeguarding goals, as well as its budgetary limitations—will need to admit the dangerous nuclear materials (e.g., direct-use materials, such as highly enriched uranium, mixed-oxide fuels, and separated plutonium) and activities (e.g., nuclear fuel-making, especially at bulk-handling facilities) for which it cannot provide timely warning of diversion, and thus cannot effectively safeguard. Moreover, the Agency should make a point of describing and identifying its accountancy, inspection, containment, and surveillance of these still unsafeguardable activities and materials as, at best, "monitoring" rather than "safeguarding."[101]

In short, with greater candor and clarity about the IAEA's safeguarding shortfalls, the Agency can help governments to clarify the line between effectively safeguardable, and therefore truly "safe," nuclear materials, technologies, and activities; and those which are not currently safeguardable and thus not merely "sensitive," but also inherently "dangerous."[102]

The Role of Legal Clarity by Governments.

Within the last few years, Iran's nuclear intransigence and North Korea's nuclear detonation have created a greater sense of urgency among governments seeking to curb nuclear proliferation. For example, in an attempt to clarify further the extent to which nuclear technology, materials, and know-how should be exchanged, the French Republic went so far as to propose a set of criteria during the lead-up to the 2005 quadrennial NPT review conference, criteria which importing states would need to meet in order to receive nuclear goods. "The export of such materials, facilities, equipment, or related technologies," France suggested in a May 2004 working paper, "should only be envisaged in the light of the existence of a set of conditions relevant to the global nonproliferation regime and NPT objectives"—conditions such as:

- an alleged energy need in the [importing] country;
- a credible nuclear power generation program and related fuel cycle needs;
- an economically rational plan for developing such projects;

- an Additional Protocol [granting the IAEA greater legal authority to inspect for undeclared nuclear materials and activities] brought into force and implemented before any physical transfer or transfer of know-how;
- the highest standard of nonproliferation commitments;
- the effective and efficient implementation of an export control system with adequate sanctions;
- the highest standard of nuclear security and safety;
- an analysis of the stability of the country and the region concerned.[103]

Within recent months, moreover, the U.S. Government has again signaled its support of proposals in the French working paper.[104]

That said, the United States and like-minded governments have yet to counter directly readings of the NPT's Article IV recognizing the per se right, or the unqualified per se right, to nuclear fuel-making. One can think of Article IV as international law's equivalent of a Rorschach Test: What a government claims to see in this treaty provision—either *de jure* cover for its approach to *de facto* status as *virtual nuclear-weapon state*, or clear criteria limiting the scope of "nuclear energy for peaceful purposes" to which signatories have an "inalienable right"—certainly reveals a great deal about how it views the NPT's fundamental and overriding goal of nuclear nonproliferation.

Certainly, the clarity or confusion with which governments seeking to curb nuclear proliferation interpret Article IV will substantially impact the decisions of other NPT signatories in the not-too-distant future. Although a consistent and sustainable reading of the NPT, by itself, cannot prevent the emergence of future proliferation problems, it can provide governments with a clear and legal foundation for effective policies which, at the very least, delegitimize unqualified per se readings of Article IV, and thus strongly discourage other NPT signatories from imitating, or even improving upon, the North Korean and Iranian examples. In contrast, a confused and muddled answer—or, equally as bad, no response at all—will have precisely the opposite effect. It will encourage ever more signatories to believe, and act on the belief, that they have a right under all circumstances, even non-compliance with NPT and IAEA obligations, to any nuclear activity short of inserting fissile material into a nuclear weapon.

In such a world, signatories in full compliance with their NPT and IAEA obligations would face, to borrow key phrases from Article 32 of the *Vienna Convention on the Law of Treaties*, the "manifestly absurd" and "unreasonable" outcome of ever more *virtual nuclear-weapon states* like Iran, and ever more actual nuclear-armed states like North Korea.

*****

The author wishes to thank Henry Sokolski, executive director of the Nonproliferation Policy Education Center (NPEC), for his generous feedback and continuing guidance. Indeed, Mr. Sokolski's article, "Nuclear Rights and Wrongs" (*National Review Online*, September 21, 2004), along with Paul Lettow's unpublished essay, "Fatal Flaw? The NPT and the Problem of Enrichment and Reprocessing" (April 2005), provided early sources of information and inspiration. All faults in this essay belong to the author, however.

The author presented an early version of this chapter as "Life, Liberty, and the Pursuit of Fissile Material? How the NPT's Preamble and Article V Can Help To Properly Interpret Articles III and IV," to "Assessing the IAEA's Ability to Safeguard Peaceful Nuclear Energy," a symposium, held on November 13, 2006, at the Centre de Conférences Internationales in Paris, France, which was co-hosted by the French Ministry of Foreign Affairs, the *Fondation pour la Recherche Stratégique* and NPEC. This symposium was attended by officials from the French, American, and German governments, as well as the IAEA. Attendees also provided feedback that helped the author to improve the chapter.

ENDNOTES - CHAPTER 8

1. After negotiations for the *Treaty on the Nonproliferation of Nuclear Weapons* concluded in 1968, a handful of analysts warned, in classified and unclassified reports in the years following, that the spread of nuclear fuel-making would encourage the emergence of many more non-nuclear-weapon states within months, or even days, of completing a nuclear explosive device. For example, see Richard Rosecrance, *After the NPT, What?* Washington, DC: U.S. Department of State, Policy Planning Council, May 28, 1968, SECRET (Declassified on December 6, 1994), Declassified Documents Reference System Document ("DDRS") No. CK3100082939; Victor Gilinsky and William Hoehn, *Nonproliferation Treaty Safeguards and the Spread of Nuclear Technology*, R-501, Santa Monica, CA: RAND Corporation, May 1970; and Albert Wohlstetter, *et al.*, *Moving Toward Life in a Nuclear Armed Crowd?* PH-76-04-389-14, final report to the U.S. Arms Control and Disarmament Agency (ACDA), Contract No. ACDA/PAB-263, Los Angeles, CA: PAN Heuristics, December 4, 1975, Revised April 22, 1976.

2. Dr. Mohamed ElBaradei, "Addressing Verification Challenges," Statement of the IAEA Director General to the *Symposium on International Safeguards*, Vienna, Austria, October 16, 2006, available from *www.iaea.org/NewsCenter/Statements/2006/ebsp2006n018.html*. The author of this chapter attended the symposium.

3. *Ibid*. ElBaradei's arguments about what he now terms *virtual nuclear-weapon state*s are similar to those that Albert Wohlstetter and colleagues made in studies on the military potential of civilian nuclear energy that were conducted for various U.S. government agencies in the 1970s. For example, as Wohlstetter and company

argued in a 1979 report to the Arms Control and Disarmament Agency:

> The interpretation of Article IV is by no means a trivial matter. If, in fact, technological transfers can bring a "nonnuclear weapon state" within weeks, days or even hours of the ability to use a nuclear explosive, [then] *in the operational sense* that "nonnuclear weapon state" will have nuclear weapons [emphasis added]. The point is even more fundamental than the fact that effective safeguards [according to the IAEA] mean timely warning. A necessary condition for timely warning is that there be a substantial elapsed time. But if there is no substantial elapsed time before a government may use nuclear weapons, [then] in effect it *has* them [emphasis original].

See Albert Wohlstetter, Gregory Jones, and Roberta Wohlstetter, *Why the Rules Have Needed Changing*, Part I of II in *Towards a New Consensus on Nuclear Technology*, Vol. 1, PH-78-04-832-33, a summary report prepared for ACDA, Contract No. AC7NC106, Los Angeles, CA: Pan Heuristics, July 6, 1979, pp. 36-37.

4. *Treaty on the Non-Proliferation of Nuclear Weapons*, July 1, 1968, entered into force on March 5, 1970, 21 U.S.T. 483, 729 UNT.S. 161. (Hereafter NPT.)

5. Kofi Annan, "Secretary-General's Address to the Nuclear Non-Proliferation Treaty Review Conference," New York, NY, May 2, 2005, available from *www.un.org/events/npt2005/statements/npt02.sg.pdf*.

6. Article I of the NPT states:

> Each nuclear-weapon State Party to the Treaty undertakes not to transfer to any recipient whatsoever nuclear weapons or other nuclear explosive devices or control over such weapons or explosive devices directly, or indirectly; and not in any way to assist, encourage, or induce any non-nuclear-weapon State to manufacture or otherwise acquire nuclear weapons or other nuclear explosive devices, or control over such weapons or explosive devices.

Article II states:

> Each non-nuclear-weapon State Party to the Treaty undertakes not to receive the transfer from any transferor whatsoever of nuclear weapons or other nuclear explosive devices or of control over such weapons or explosive devices directly, or indirectly; not to manufacture or otherwise acquire nuclear weapons or other nuclear explosive devices; and not to seek or receive any assistance in the manufacture of nuclear weapons or other nuclear explosive devices.

7. Article IV of the NPT states:

> 1. Treaty to develop research, production and use of nuclear energy for peaceful purposes without discrimination and in conformity with Articles I and II of this Treaty.
>
> 2. All the Parties to the Treaty undertake to facilitate, and have the right to participate in, the fullest possible exchange of equipment, materials and scientific and technological information for the peaceful uses of nuclear energy. Parties to the Treaty in a position to do so shall also co-operate in contributing alone or together with other States or international organizations to the further development of the applications of nuclear energy for peaceful purposes, especially in the territories of non-nuclear-weapon States Party to the Treaty, with due consideration for the needs of the developing areas of the world.

8. For example, a number of Non-Aligned Movement (NAM) governments that are signatories of the NPT argued in the 2006 NAM summit's nonlegally-binding final document that: "each country's choices and decision in the field of peaceful uses of nuclear energy should be respected without jeopardizing its policies or international cooperation agreements and arrangements for peaceful uses of nuclear energy and its fuel-cycle policies." See Fourteenth Summit Conference of the Heads of State or Government of the Non-Aligned Movement, Final Document, NAM 2006/Doc.1/Rev. 3, September, 16, 2006, para. 95, available from *www.cubanoal.cu/ingles/docadoptados/docfinal.htm.*

9. See IAEA, *Implementation of the NPT Safeguards Agreement in the Islamic Republic of Iran*, Report by the Director General, GOV/2003/40, June 6, 2003, *esp.* paras. 32-35, available from *www.iaea.org/Publications/Documents/Board/2003/gov2003-40.pdf*; *Implementation of the NPT Safeguards Agreement in the Islamic Republic of Iran*, Report by the Director General, GOV/2003/63, August 26, 2003, *esp.* paras. 47-53, available from *www.iaea.org/Publications/Documents/Board/2003/gov2003-63.pdf*; *Implementation of the NPT Safeguards Agreement in the Islamic Republic of Iran*, Report by the Director General, GOV/2003/75, November 10, 2003, *esp.* paras. 45-56, available from *www.iaea.org/Publications/Documents/Board/2003/gov2003-75.pdf*.

10. See IAEA, *Implementation of the NPT Safeguards Agreement in the Islamic Republic of Iran*, Report by the Director General, GOV/2004/83, November 15, 2004, *esp.* paras. 85-114, available from *www.iaea.org/Publications/Documents/Board/2004/gov2004-83.pdf*; and *Implementation of the NPT Safeguards Agreement in the Islamic Republic of Iran*, Report by the Director General, GOV/2005/67, September 2, 2005, *esp.* paras. 42-52, available from *www.iaea.org/Publications/Documents/Board/2005/gov2005-67.pdf*.

11. See IAEA, *Implementation of the NPT Safeguards Agreement in the Islamic Republic of Iran*, Resolution Adopted by the IAEA Board of Governors, GOV/2005/77, September 24, 2005, available from *www.iaea.org/Publications/Documents/Board/2005/gov2005-77.pdf*.

12. See *ibid*, and Security Council of the United Nations, *Resolution 1696*, S/RES/1696, July 31, 2006, available from *www.un.org/Docs/sc/unsc_resolutions06.htm*, *Resolution 1737*, S/RES/1737, December 27, 2006, available from *www.un.org/Docs/sc/unsc_resolutions06.htm*, and *Resolution* 1747, S/RES/1747, March 24, 2007, available from *www.un.org/Docs/sc/unsc_resolutions07.htm*.

13. In early 2006, for example, the Iranian ambassador to the IAEA argued before the Agency's Board of Governors that his government possessed an "inalienable right [to] peaceful uses of nuclear energy, including nuclear fuel cycle and research and development, as envisaged in the Agency's Statute and the NPT."

See "Statement by Ali Ashgar Soltanieh, Resident Representative of the Islamic Republic of Iran to the IAEA," Vienna, Austria, February 2, 2006, p. 10, available from *www.iaea.org/Publications/Documents/Infcircs/2006/infcirc666.pdf*. Moreover, when the Iranian government rejected a mid-2006 proposal from the five permanent members of the Security Council and Germany to abandon enrichment and other nuclear fuel-making activities in exchange for political, economic, and technological incentives, it asserted: "Development of [Iran's] peaceful nuclear program is based on its *specific and undeniable rights* [emphasis added] under the NPT. [Iran] cannot accept deprivation from its legal rights in [the] development and use of peaceful nuclear energy including the fuel cycle, and continuing research and development of enrichment process as underscored in the NPT and IAEA safeguards." See Islamic Republic of Iran, *Response to the Package Presented on June 6, 2006*, English trans., published by the Council on Foreign Relations on August 22, 2006, available from *www.cfr.org/publication/11432/*.

14. Here it is worth noting that individual members of the U.S. Government's legislative and executive branches have attempted to counter readings of the NPT that affirm a per se right, or unqualified per se right, of signatories to enrichment and reprocessing, ENR) and other sensitive nuclear fuel-making technologies. For example, see Senators Richard Lugar (R-IN) and Evan Bayh (D-IN), "A Nuclear Fuel Bank Advocated," op-ed, *Chicago Tribune*, October 22, 2006. "For too long, the Nuclear Non-Proliferation Treaty has been exploited," Lugar and Bayh wrote. "We need a new international non-proliferation standard that prevents countries from using the guise of nuclear energy to develop nuclear weapons," they added. "The dangers are so great that the world community must declare that there is no right under the Nuclear Non-Proliferation Treaty to enrich uranium or separate plutonium from spent nuclear fuel." See *also* John R. Bolton, "The NPT: A Crisis of Non-Compliance," Statement of the Under Secretary of State for Arms Control and International Security to the Third Session of the Preparatory Committee for the 2005 Review Conference of the Treaty on the Non-Proliferation of Nuclear Weapons, New York, April 27, 2004, available from *www.state.gov/t/us/rm/31848.htm*. With respect to ENR technologies, Bolton asserted, "The Treaty provides no right to such sensitive fuel cycle technologies." Most recently, Congresswoman Ileana Ros-Lehtinen (R-FL), the ranking member of the House Committee

on Foreign Affairs, argued against reading the NPT's Article IV as recognizing an "absolute right" to nuclear fuel-making. See Ileana Ros-Lehtinen, "The Right to Survive: Nuclear 'Rights" Don't Trump It," *National Review Online*, October 5, 2007, available from *article.nationalreview.com/?q=MDYyNjBjNWM3MzBlZGRlYjBjOD M3N2ZkZDcxOGViODk=*.

15. NPT, preamble, para. 6.

16. NPT, Art. III, para. 1.

17. IAEA, *The Structure and Content of Agreements Between the Agency and States Required in Connection with the Treaty on the Non-Proliferation of Nuclear Weapons*, INFCIRC/153, Corrected, June 1972, para. 28. In February 1992, after the IAEA had discovered the extent to which Ba'athist Iraq had failed to declare a host of nuclear materials and activities, the Agency's Board of Governors affirmed that INFCIRC/153's objective of detecting and deterring the diversion of nuclear material applies not only "to nuclear material declared by a State," but also to "any nuclear material [and related activities] subject to safeguards that *should have been declared*." See IAEA, *The Safeguards System of the International Atomic Energy Agency*, undated, circa 2002, para. 13, available from *www.iaea.org/OurWork/SV/Safeguards/safeg_system.pdf*.

18. INFCIRC/153, Corrected, para. 29.

19. For background on SAGSI, see Marvin M. Miller, "Are IAEA Safeguards on Plutonium Bulk-Handling Facilities Effective?" Washington, DC: Nuclear Control Institute, August 1990, available from *www.nci.org/k-m/mmsgrds.htm*. As Miller notes, "The values recommended by SAGSI for the detection goals were carefully described as provisional guidelines for inspection planning and for the evaluation of safeguards implementation, not as requirements, and were so accepted by the Agency."

20. *IAEA Safeguards Glossary*, 2001 Edition, International Nuclear Verification Series No. 3, Vienna, Austria: IAEA, June 2002, sec. 3, para. 14, available from *www-pub.iaea.org/MTCD/publications/PDF/nvs-3-cd/PDF/NVS3_prn.pdf*.

21. See Thomas B. Cochran, "Adequacy of IAEA's Safeguards for Achieving Timely Detection," an essay presented at "After Iran: Safeguarding Peaceful Nuclear Energy," London, UK, October 2005 p. 2, available from *www.npec-web.org/Frameset.asp?PageType=Single&PDFFile=Paper050930CochranAdequacyofTime&PDFFolder=Essays*. See also Cochran and Christopher E. Paine, "The Amount of Plutonium and Highly-Enriched Uranium Needed for Pure Fission Nuclear Weapons," New York: Natural Resources Defense Council, August 22, 1994, Revised April 13, 1995, available from *www.nrdc.org/nuclear/fissionw/fissionweapons.pdf*.

22. *IAEA Safeguards Glossary*, sec. 3, para. 13.

23. As Marvin Miller of the Massachusetts Institute of Technology argued in August 1990:

> [Timely warning of diversion] is taken to be detection of a diversion quickly enough to take diplomatic action to prevent the fabrication and insertion of the diverted material into a first bomb that is otherwise complete. Thus, detection time must be even shorter than conversion time, in order to allow for evaluation and response.

See Marvin M. Miller, "Are IAEA Safeguards on Plutonium Bulk-Handling Facilities Effective?" See *also* Paul Leventhal, "Safeguards Shortcomings: A Critique," Washington, DC: Nuclear Control Institute, September 12, 1994, available from *www.nci.org/p/plsgrds.htm*; and Thomas B. Cochran, "Adequacy of IAEA's Safeguards for Achieving Timely Detection," pp. 5-12.

24. *IAEA Safeguards Glossary*, sec. 3, para. 13, table I.

25. *Ibid.*, sec. 3, para. 20.

26. See D. E. Ferguson, "Simple, Quick Processing Plant," Memorandum to F. L. Culler, Oak Ridge National Laboratory, August 30, 1977. See *also* J. A. Hassberger, "Light-Water Reactor Fueling Handling and Spent Fuel Characteristics," Fission Energy and Systems Safety Program, Lawrence Livermore National Laboratory, circa February 26, 1999.

27. As Gilinsky, Miller and Hubbard argued:

> Having a gas centrifuge plants producing LEU makes it much easier to construct and operate a clandestine one. The presence of the larger plant would mask many of the intelligence indicators and environmental indications of a clandestine one so it would harder to find.
>
> But even in the absence of any commercial enrichment—in the case of a country with one or more stand alone LWRs—the presence of LWRs means that a substantial supply of fresh LWR fuel would also be present at times. That such fresh fuel can provide a source of uranium for clandestine enrichment is another possibility that has received essentially no attention in proliferation writings. Since the fuel is already low enriched uranium, a much smaller gas centrifuge plant would suffice to raise the enrichment to bomb levels than would be the case if the starting point is natural uranium. By starting with such LEU fuel pellets, which are uranium oxide, UO2, the enricher would be able to skip the first five processes required to go from uranium ore to uranium hexafluoride gas, the material on which the gas centrifuge operate. To go from the uranium oxide pellets to uranium hexafluoride the would-be bomb-maker would crush the pellets and react the powder with fluorine gas. Suitably processed, the LEU pellets could provide feed for clandestine enrichment.

See Gilinsky *et al.*, *A Fresh Examination of the Proliferation Dangers of Light Water Reactors*, Washington, DC: Nonproliferation Policy Education Center, October 22, 2004, p. 14, available from *www.npec-web.org/Frameset.asp?PageType=Single&PDFFile=20041022-GilinskyEtAl-LWR&PDFFolder=Essays*.

28. *Ibid*.

29. *IAEA Safeguards Glossary*, sec. 3, para. 22.

30. The *IAEA Safeguards Glossary* more technically defines a "material balance period" as:

> Under an INFCIRC/153-type safeguards agreement, the term is used to refer to the time between two consecutive physical inventory takings (PITs) as reflected in the State's material balance report. Under an INFCIRC/66-type safeguards agreement, the term is used to refer to what more accurately should be called the book balance period, since the beginning and the ending dates of the period are not necessarily linked to PITs.

See *ibid.*, sec. 6, para. 47.

31. *Ibid.*, sec. 3, para. 10.

32. *Ibid.*, sec. 3, para. 24, emphasis added.

33. *Ibid.*, sec. 3, para. 23, emphasis added.

34. *Ibid.*, sec. 5, para. 28.

35. *Ibid.*, sec. 3, para. 17.

36. As Marvin M. Miller explained in 1990:

> A relevant example is the planned 800 tonne/yr Rokkasho reprocessing facility at Aomori in Japan. Assuming that: (1) the plant processes spent fuel with an average total plutonium content of 0.9%, (2) the error in measuring the MUF [material unaccounted for], specified by (MUF), is dominated by the error in measuring the plutonium input, and is equal to 1% of this input, and, (3) the material balance calculation is done once a year, then the absolute value of (MUF) = 72 kg of Pu/yr. It is straightforward to show that the minimum amount of diverted plutonium which could be distinguished from this measurement "noise" with detection and false alarm probabilities of 95% and 5%, respectively, is 3.3 (MUF, or 246 kg in this example, equivalent to more than 30 significant quantities.

See Miller, "Are IAEA Safeguards on Plutonium Bulk-Handling Facilities Effective?"

37. Paul Leventhal, "Safeguards Shortcomings: A Critique."

38. Pierre Goldschmidt, "The Nuclear Non-proliferation Regime: Avoiding the Void," Washington, DC: Nonproliferation Policy Education Center, February 28, 2006, p. 3, available from *www.npec-web.org/Frameset.asp?PageType=Single&PDFFile=Paper060221%20Goldschmidt%20-%20The%20Nuclear%20Non-proliferation%20Regime%20-%20Avoiding%20the%20Void&PDFFolder=Essays.*

39. Edwin S. Lyman, "Can Nuclear Fuel Production in Iran and Elsewhere Be Safeguarded Against Diversion?" an essay presented at "After Iran: Safeguarding Peaceful Nuclear Energy," London, UK, October 2005, p. 7, available from *www.npec-web.org/Frameset.asp?PageType=Single&PDFFile=Paper050928LymanFuelSafeguardDiv&PDFFolder=Essays.*

40. *Ibid.*, p. 15.

41. For background on the issue of IAEA resources, see Thomas Shea, "Financing IAEA Verification of the NPT," a paper presented to "Assessing the IAEA's Ability to Verify the NPT," Paris, France, November 12-13, 2006, available from *www.npec-web.org/Essays/20061113-Shea-FinancingIAEAVerification.pdf.*

42. Table 3 is adapted from Henry D. Sokolski, "Clarifying and Enforcing the Nuclear Rules," prepared testimony before "Weapons of Mass Destruction: Current Nuclear Proliferation Challenges," a hearing before the Committee on Government Reform's Subcommittee on National Security, Emerging Threats, and International Relations, U.S. House of Representatives, September 6, 2006 p. 3, note 2, available from *www.npec-web.org/Frameset.asp?PageType=Single&PDFFile=20060921-FINAL-Sokolski-TestimonyHouseSubcommittee&PDFFolder=Testimonies.*

43. ElBaradei, "Addressing Verification Challenges."

44. Fred C. Iklé, *How Nations Negotiate,* New York Praeger, 1964, p. 15.

45. *Ibid.*

46. Vienna Convention on the Law of Treaties, May 23, 1969, entered into force on January 27, 1980, 1155 UNT.S. 331 (hereafter VCLT). Although the U.S. Government is not a party to the VCLT, the *Restatement (Third) of Foreign Relations Law of the United States* notes that the VCLT "represents generally accepted principles and the United States has also appeared willing to accept them despite differences of nuance and emphasis." See *Restatement (Third) of Foreign Relations Law of the United States*, Washington, DC: American Law Institute, 1987, Part III, Ch. 3, Sec. 325, Comment a.

47. *Ibid.*, Art. 31, para. 1.

48. *Ibid.*, Art. 32.

49. There are a number of histories of the negotiations that led eventually to the NPT. For examples of unclassified histories, see U.S. Arms Control and Disarmament Agency, *International Negotiations on the Treaty on the Nonproliferation of Nuclear Weapons*, Publication No. 48, Washington, DC: U.S. Government Printing Office, January 1969; Mohamed Ibrahim Shaker, *The Treaty on the Non-Proliferation of Nuclear Weapons: A Study Based on the Five Principles of UN General Assembly Resolution 2028, XX*, Doctoral Dissertation, Geneva, Switzerland: University of Geneva, Department of Political Science, 1976; Arthur Steiner, "Article IV and the 'Straightforward Bargain'," PAN Heuristics Paper 78-832-08, in Wohlstetter, *et al.*, *Towards a New Consensus on Nuclear Technology*, Vol. II, Supporting Papers, ACDA Report No. PH-78-04-832-33, Marina del Rey, CA: Pan Heuristics, 1979, pp. 1-8; Eldon V. C. Greenberg, "NPT and Plutonium: Application of NPT Prohibitions to 'Civilian' Nuclear Equipment, Technology, and Materials Associated with Reprocessing and Plutonium Use," Washington, DC: Nuclear Control Institute, 1984, Revised May 1993; Henry D. Sokolski, *Best of Intentions: America's Campaign Against Strategic Weapons*, Westport, CT: Praeger, 2001; and Paul Lettow, "Fatal Flaw? The NPT and the Problem of Enrichment and Reprocessing," unpublished essay, April 27, 2005. In addition, a number of declassified histories of the NPT negotiations, which were written from the institutional viewpoints of various U.S. Government agencies, are now available. For example, see U.S. Arms Control and Disarmament Agency, *The U.S. Arms Control and Disarmament Agency during the Johnson Administration*, Volume

I: "Summary and Analysis of Principal Developments," undated, circa 1969, CONFIDENTIAL, Declassified on August 21, 1992, DDRS No. CK3100005560; ACDA, *The U.S. Arms Control and Disarmament Agency during the Johnson Administration*, Volume II: "Policy and Negotiations," Part B: "Non-Proliferation of Nuclear Weapons," January 2, 1969, SECRET/NOFORN, Declassified on March 22, 1999, DDRS No. CK3100152990; and U.S. Atomic Energy Commission, *The Atomic Energy Commission during the Administration of Lyndon B. Johnson, November 1963-January 1969,* Volume I: "Administrative History," Part II, undated, circa 1969, SECRET, Declassified May 10, 1994, DDRS No. CK3100062802, *esp.* Chapter 9, "Negotiation of the Nonproliferation Treaty."

50. The Eighteen Nation Disarmament Committee (ENDC) grew out of the so-called Ten Nation Committee on Disarmament (TNCD). The TNCD was formed in late 1959 through an agreement among the United States, USSR, Britain, and France. It consisted of five nations from the West: the United States, Britain, France, Canada, and Italy; and five nations from the Soviet bloc: the USSR, Bulgaria, Czechoslovakia, Poland, and Romania. See "Four-Power Communiqué on Disarmament Negotiations," September 7, 1959, in U.S. Department of State, Bureau of Public Affairs, Historical Office, *Documents on Disarmament, 1945-1959*, Vol. 2 of 2, 1957-1959, Publication No. 7008, Washington, DC: U.S. Government Printing Office, August 1960, pp. 1441-1442. In late 1960, after negotiations in the TNCD had stalled, the USSR proposed adding five states—namely, Ghana, India, Indonesia, Mexico, and the United Arab Republic—to the TNCD. The United States, however, did not endorse the proposal, and instead criticized the USSR for walking out of TNCD negotiations earlier in the year. See "Soviet Draft Resolution Submitted to the General Assembly: Enlargement of the Ten Nation Committee on Disarmament," September 26, 1960, in U.S. Department of State, Bureau of Public Affairs, Historical Office, *Documents on Disarmament, 1960*, Publication No. 7172, Washington, DC: U.S. Government Printing Office, July 1961, pp. 250-251, and p. 250, note 1. In mid-1961, though, the United States proposed to the USSR that the TNCD be expanded to include 20 members. See "United States Memorandum Submitted During the Bilateral Talks with the Soviet Union: Composition of the Disarmament Forum," July 29, 1961, in U.S. Arms Control and Disarmament Agency, *Documents on Disarmament, 1961*, Publication No. 5, Washington, DC: U.S.

Government Printing Office, August 1962, pp. 271-273. The Soviet Union initially rejected this proposal, however. See "Statement by the Soviet Government on the Bilateral Talks," September 22, 1961, *ibid.*, pp. 444-458. In mid-to-late 1961, the United States nevertheless pushed this proposal in the UN General Assembly. See "Statement by the United States Representative (Stevenson) to the First Committee of the General Assembly," November 15, 1961, in *ibid.*, pp. 616-631.

51. By late 1961, the United States and USSR were able to agree to an expanded Eighteen Nation Disarmament Committee, which would include the original 10 members of the Ten Nation Committee on Disarmament, plus eight nonaligned members—namely, Brazil, Burma, Ethiopia, India, Mexico, Nigeria, Sweden, and the United Arab Republic. The UN General Assembly endorsed the proposed ENDC with a resolution. See "General Assembly Resolution 1722 (XVI): Question of Disarmament," December 20, 1961, available from *www.un.org/documents/ga/res/16/ares16.htm*.

52. UN General Assembly Resolution 1722 (XVI) endorsed the formation of ENDC, and recommended that it resume negotiations on an agreement for "general and complete disarmament under effective international control." See "General Assembly Resolution 1722 (XVI): Question of Disarmament," December 20, 1961, in *ibid.*

53. At the opening of the ENDC's eighth session, the United States called for delegations to focus instead on concluding a nuclear nonproliferation treaty consistent with UN General Assembly Resolution 1665 (XVI) also known as "the Irish Resolution." See "Statement by ACDA Director Foster to the Eighteen Nation Disarmament Committee," ENDC/PV. 218, July 27, 1965, in U.S. Arms Control and Disarmament Agency, *Documents on Disarmament, 1965*, Publication No. 34, Washington, DC: U.S. Government Printing Office, December 1966, pp. 281-286. For the Irish Resolution's text, see "General Assembly Resolution 1665 (XVI): Prevention of the Wider Dissemination of Nuclear Weapons," December 4, 1961, available from *www.un.org/documents/ga/res/16/ares16.htm*. For background on how, in the aftermath of Communist China's 1964 nuclear detonation, the U.S. Government came to view the conclusion of a nuclear

nonproliferation treaty as an urgent priority, see "Report by the Committee on Nuclear Proliferation" (Gilpatric Committee), January 21, 1965, in U.S. Department of State, *Foreign Relations of the United States,* Vol. 11, No. 64, Washington, DC: U.S. Government Printing Office, 1997, available from *www.state.gov/www/about_state/history/vol_xi/g.html*. For a more recent historical account of the Gilpatric Committee's role in changing the Johnson Administration's nonproliferation policy, see Frank J. Gavin, "Blasts from the Past: Proliferation Lessons from the 1960s," *International Security,* Vol. 29, No. 3, Winter 2004/05, pp. 112-113.

54. *Treaty for the Prohibition of Nuclear Weapons in Latin America and the Caribbean,* February 14, 1967, entered into force April 22, 1968, 634 UNT.S. 326, Art. 17. (Hereafter Treaty of Tlatelolco.)

55. "From U.S. Embassy, Mexico, Telegram No. 2419," U.S. Department of State, February 13, 1967, CONFIDENTIAL, Declassified on January 3, 1980, DDRS Document No. CK3100432606.

56. "Redraft of Possible Non-Proliferation Treaty Formations," U.S. Department of State, April 17, 1967, SECRET, Declassified on April 14, 1999, DDRS Document No. CK3100472688.

57. U.S. Arms Control and Disarmament Agency, *The U.S. Arms Control and Disarmament Agency during the Johnson Administration,* Volume I: "Summary and Analysis of Principal Developments," undated, circa 1969, CONFIDENTIAL, Declassified on August 21, 1992, DDRS No. CK3100005560, p. 56.

58. "Draft Treaty on the Nonproliferation of Nuclear Weapons," ENDC/192 (U.S. submission) and ENDC/193 (USSR submission), August 24, 1967, in *U.S. Arms Control and Disarmament Agency, Documents on Disarmament, 1967,* Publication No. 46, Washington, DC: U.S. Government Printing Office, July 1968, pp. 394-395.

59. Mexico's working paper proposed the following formulation for Article IV:

> 1. Nothing in this Treaty shall be interpreted as affecting the inalienable right of all the Parties to the Treaty to develop research, production, and use of nuclear energy

for peaceful purposes without discrimination and in conformity with Articles I and II of this Treaty.

2. All the Parties to this Treaty have the right to participate in the fullest possible exchange of scientific and technological information on the peaceful uses of nuclear energy. Those Parties that are in a position to do so, have *the duty* to contribute, according to their ability, alone or in cooperation with other States or international organizations, to the further development of the production, industries, and other applications of nuclear energy for peaceful purposes, especially in the territories of non-nuclear-weapon States (emphasis added).

See "Mexican Working Paper Submitted to the Eighteen Nation Disarmament Committee: Suggested Additions to Draft Nonproliferation Treaty," ENDC/196, September 19, 1967, in U.S. Arms Control and Disarmament Agency, *Documents on Disarmament, 1967*, Publication No. 46, Washington, DC: U.S. Government Printing Office, July 1968, pp. 394-395. Mexico's working paper was cited by Eldon V. C. Greenberg, "The NPT and Plutonium: Applications of NPT Prohibitions to 'Civilian' Nuclear Equipment, Technology, and Materials Associated with Reprocessing and Plutonium Use," Washington, DC: Nuclear Control Institute, 1984, Revised May 1993, p. 16 and note 63, available from *www.nci.org/03NCI/12/NPTandPlutonium.pdf.*

60. Romania's working paper proposed the inclusion of the following language in the treaty's preamble: "Affirming *the absolute right* of all States, whether they possess nuclear weapons or not, to undertake research on the peaceful applications of nuclear energy and to use nuclear energy for peaceful purposes, both now and in the future, on the basis of equality and without any discrimination" (emphasis added). See "Romanian Working Paper Submitted to the Eighteen Nation Disarmament Committee: Amendments and Additions to the Draft Nonproliferation Treaty," ENDC/199, October 19, 1967, in U.S. Arms Control and Disarmament Agency, *Documents on Disarmament, 1967*, pp. 525-526.

61. Brazil's proposed amendments suggested that Article IV contain the following language:

> Nothing in this Treaty shall be interpreted as affecting the inalienable right of all the Parties to the Treaty to develop, alone or in cooperation with other States, research, production, and use of nuclear energy for peaceful purposes, *including nuclear explosive devices for civil uses*, without discrimination, as well as the right of the Parties to participate in the fullest possible exchange of information for, and to contribute or in cooperation with other States to, the further development of the applications of nuclear energy for peaceful purposes (emphasis added).

See "Brazilian Amendments to the Draft Nonproliferation Treaty," ENDC/201, October 31, 1967, in *ibid.*, p. 546.

62. In early November 1967, the Nigerian delegation proposed the following amendments to Article IV:

> ARTICLE IV A:
>
> Each Party to the Treaty undertakes to cooperate directly or through the IAEA, in good faith and according to its technological and/or material resources, with any other State or group of States Party to this Treaty in the development and advancement of nuclear technology for peaceful purposes, and in the fullest possible exchange of scientific and technological information on the peaceful uses of nuclear energy.
>
> The nuclear weapon States Party to this Treaty shall make available through the IAEA, to all non-nuclear weapon Parties, full scientific and technological information on the peaceful applications of nuclear energy accruing from research on nuclear explosive devices.
>
> The nuclear weapon States Party to the Treaty shall also provide facilities for scientists from non-nuclear weapon countries Party to the Treaty to collaborate with their scientists working on nuclear explosive devices, in order to narrow the intellectual gap which will be created in that field as a result of restrictions imposed by this Treaty on non-nuclear weapon States.

> ARTICLE IV B:
>
> Each Party to the Treaty undertakes to communicate annually to the IAEA, full information on the nature, extent and results of its cooperation with any other Party or group of Parties, in the development of nuclear energy for peaceful purposes. The Reports so received by the IAEA shall be circulated by the Agency to all the Parties to the Treaty.
>
> ARTICLE IV C:
>
> Each Party to this Treaty shall take necessary legal and administrative steps to ensure that all organisations working on the development of nuclear energy in territory under its jurisdiction do so conformity with the aims and provisions of the Treaty.

See "Nigerian Working Paper Submitted to the Eighteen Nation Disarmament Committee: Additions and Amendments to the Draft Nonproliferation Treaty," ENDC/202, November 2, 1967, in *ibid.*, p. 558.

63. "Spanish Memorandum to the Co-Chairman of the ENDC," ENDC/210, February 8, 1968, in U.S. Arms Control and Disarmament Agency, *Documents on Disarmament, 1968*, Publication No. 52, Washington, DC: U.S. Government Printing Office, September 1969, p. 40. Spain's memorandum was cited by Greenberg, "The NPT and Plutonium," p. 16 and note 64.

64. Brazil proposed that Article IV be revised to read:

> Nothing in this Treaty shall be interpreted as affecting the inalienable right of all the Parties to the Treaty to develop, along or in cooperation with other States, research, production and use of nuclear energy for peaceful purposes, *including nuclear explosive devices for civil uses*, without discrimination (emphasis added).

See "Brazilian Amendments to the Draft Treaty on Nonproliferation of Nuclear Weapons," ENDC/201/Rev. 2, February 13, 1968, in U.S. Arms Control and Disarmament Agency, *Documents on Disarmament, 1968*, p. 64.

65. Italian Working Paper Submitted to the Eighteen Nation Disarmament Committee: Additions and Amendments to Articles IV, VIII, and X of the Draft Nonproliferation Treaty, ENDC/218, February 20, 1968, Corr. 2, February 22, 1968, in *ibid.*, p. 92.

66. Moreover, though the NPT views "nuclear energy for peaceful purposes" through the prism of "inalienable rights," it is worth noting that an "inalienable right" is not an undeniable right. For example, though the *Declaration of Independence* affirms "life, liberty, and the pursuit of happiness" as inalienable rights of man, this affirmation does not imply that the U.S. Government cannot fine, imprison, or even execute an individual if s/he violates certain laws.

67. NPT, Art. III, para. 1.

68. *Ibid.*, Art. III, para. 2.

69. *Ibid.*, Art. III, para. 3.

70. *Ibid.*, Art. IV, para. 1.

71. *Ibid.*, Art. III, para. 1, .

72. The NPT's negotiation history confirms that the "benefits-without-discrimination" principle is meant to be understood in relation to Article IV. In August 1967, when American delegate William Foster, along with Soviet delegate Alexey Roshchin introduced the first version of what we know today as Article IV, he said that the article's two paragraphs "are *specific elaborations of the principle stated in the preamble* 'that the benefits of peaceful applications of nuclear technology should be available for peaceful purposes to all Parties . . . whether nuclear-weapon or non-nuclear-weapon States'." See "Statement by ACDA Director [William] Foster to the Eighteen Nation Disarmament Committee: Draft Nonproliferation Treaty," ENDC/PV. 325, August 24, 1967, in U.S. Arms Control and Disarmament Agency, *Documents on Disarmament, 1967*, p. 345, emphasis added.

73. NPT, preamble, para. 7.

74. *Ibid.*, Art. I.

75. *Ibid.*, Art. II.

76. Albert Wohlstetter and colleagues stressed this point in studies conducted for a number of U.S. Government agencies in the 1970s. In a 1979 report for ACDA, for example, they argued, "Article IV explicitly states that the inalienable right of all parties to the Treaty to the peaceful use of nuclear energy has to be in conformity with Articles I and II . . . [T]hese Articles are what make the Treaty a treaty against proliferation." See Wohlstetter *et al.*, *Towards a New Consensus on Nuclear Technology*, 1979, pp. 34-35. A few years later, Eldon V. C. Greenberg reiterated this point in a legal memorandum on the NPT's relation to plutonium fuel-making and use:

> There is, in short, a dynamic tension in the Treaty between its prohibitions and its injunctions to cooperate in peaceful uses of nuclear energy. An analysis of the language and history of the NPT, and particularly the key phrase "in conformity with [articles I and II]" in Article IV, paragraph 1, which is the link between the Treaty's promises and its prohibitions, tends to support the conclusion that Articles I, II and IV must be read together in such a way that assistance or activities which are ostensibly peaceful and civilian in nature do not as a practical matter lead to proliferation of nuclear weapons. The NPT, in other words, can and should be read as permitting the evaluation of such factors as proliferation risk, economic or technical justification, and safeguards effectiveness in assessing the consistency of specific or generic types of assistance and activities with the Treaty's restrictions, to ensure that action is not taken in the guise of peaceful applications of nuclear energy under Article IV which in fact is violative of the prohibitions of Articles I and II.

See "The NPT and Plutonium," p. 10.

77. NPT, Art. III, para. 1.

78. It is of interest to note that both the 2000 NPT Review Conference and 2006 summit of the Non-Aligned Movement released a politically-binding final documents affirming that "nothing in the Treaty shall be interpreted as affecting the inalienable right of all the parties to the Treaty to develop research, production and use of nuclear energy for peaceful purposes without discrimination and in conformity with articles I, II *and* III of the Treaty." See 2000 Review Conference of the Parties to the Treaty on the Non-Proliferation of Nuclear Weapons, *Final Document,* NPT/CONF.2000/28, Parts I and II, p. 8, para. 2, available from *disarmament2.un.org/wmd/npt/finaldoc.html;* and Fourteenth Summit Conference of the Heads of State or Government of the Non-Aligned Movement, *Final Document,* NAM 2006/Doc.1/Rev.3, September, 16, 2006, para. 95. I am grateful to Andreas Persbo, a nuclear law and policy researcher at the Verification Research, Training and Information Centre (VERTIC) in London, who first drew my attention to this passage of the 2000 NPT Review Conference's *Final Document.*

79. Henry D. Sokolski and George Perkovich, "It's Called *Non*proliferation," *Wall Street Journal,* Apr il 29, 2005, p. A16, available from *www.npec-web.org/Frameset.asp?PageType=Single&PDFFile=OpEd50429ItsCalledNonprolife&PDFFolder=OpEds;* and Perkovich, "Defining Iran's Nuclear Rights," Washington, DC: Carnegie Endowment for International Peace, September 7, 2006, available from *www.carnegieendowment.org/npp/publications/index.cfm?fa=view&id=18687.*

80. NPT, Art. V.

81. Exceptions included Brazil, the government of which had argued before the ENDC that non-nuclear-weapon states should be allowed to develop peaceful nuclear explosive devices. For example, see "Brazilian Amendments to the Draft Nonproliferation Treaty," ENDC/201, October 31, 1967, U.S. Arms Control and Disarmament Agency, *Documents on Disarmament, 1967,* Publication No. 46, Washington, DC: U.S. Government Printing Office, July 1968, p. 546.

82. ENDC/PV. 359, January 25, 1968, p. 6, available from *www.hti.umich.edu/e/endc/.*

83. ENDC/PV. 360, January 30, 1968, p. 7, available from *www.hti.umich.edu/e/endc/.*

84. On the comparative economics and proliferation dangers of PNEs, see Thomas Blau, *Rational Policy-Making and Peaceful Nuclear Explosives*, Doctoral Dissertation, Chicago, IL: University of Chicago, Department of Political Science, December 1972.

85. *Report of Main Committee III*, Treaty on the Nonproliferation of Nuclear Weapons Review and Extension Conference, May 5, 1995, NPT/CONF.1995/MC.III/1, sec. I, para. 2, available from *www.un.org/Depts/ddar/nptconf/162.htm.*

86. *Comprehensive Nuclear-Test-Ban Treaty*, September 10, 1996, 35 I.L.M. 1430.

87. Here I use what I view to be Article IV's three qualifications on the "nuclear energy for peaceful purposes" to which NPT signatories have an "inalienable right" to elaborate Eldon V. C. Greenberg's argument for the specific facts and circumstances under which peaceful nuclear energy should not be permissible:

> If the [proliferation] risks are great, if there can be no reasonable civilian justification for particular forms of assistance or activities, and if there can be no certainty that safeguards would be effective with respect to such assistance or activities, then a presumption should arise under the Treaty that such assistance or activities are not for permissible, peaceful purpose but are rather for a weapons or explosive purpose and therefore in violation of Articles I and II. Only in this way can there be any assurance that the NPT's objectives will be achieved.

See Greenberg, "The NPT and Plutonium," p. 13.

88. Eldon V. C. Greenberg's legal memorandum on the NPT and plutonium fuel making argues:

> [T]he distinction between permissible and impermissible activities must come down ultimately to quite pragmatic considerations. Activities must not be free from the

> Treaty's prohibitions just by virtue of being denominated "peaceful," civilian," power" or "research." The Treaty must be interpreted as viewing proliferation through something more than an "explosive lens." Rather, depending upon the facts and circumstances, assistance and activities relating to declared "peaceful," civilian," "power" or "research" purposes may be subject to the NPT's restrictions, if an evaluation of all the facts and circumstances, including such factors as economic or technical justification or effectiveness of safeguards, would indicate that the legitimacy of the assistance and/or activity is questionable. Such a pragmatic, rather than a formalistic reading of the Treaty, is most consistent with the overriding purpose of stemming the proliferation of nuclear weapons.

See Greenberg, "The NPT and Plutonium," p. 24.

89. ENDC/PV. 82, September 7, 1962, p. 37, available from *www.hti.umich.edu/e/endc/*. See also "British Paper Submitted to the Eighteen Nation Disarmament Committee: Technical Possibility of International Control of Fissile Material Production," ENDC/60, August 31, 1962, Corr. 1, November 27, 1962, in U.S. Arms Control and Disarmament Agency, *Documents on Disarmament, 1962*, Publication No. 19, Vol. 2 of 2, Washington, DC: U.S. Government Printing Office, November 1963, pp. 834-852.

90. "Statement by the Swedish Representative [Alva Myrdal] to the Eighteen Nation Disarmament Committee: Nonproliferation of Nuclear Weapons," ENDC/PV. 243, February 24, 1966, in U.S. Arms Control and Disarmament Agency, *Documents on Disarmament, 1966*, Publication No. 43, Washington, DC: U.S. Government Printing Office, September 1967, p. 56.

91. ENDC/PV. 250, March 22, 1966, p. 28, available from *www.hti.umich.edu/e/endc/*.

92. NPT, Art. IV, para. 2.

93. As strategist Albert Wohlstetter and colleagues argued in a 1979 report for the U.S. Arms Control and Disarmament Agency:

> If the "fullest possible exchange" were taken to include the provision of stocks of highly concentrated fissile material within days or hours of being ready for incorporation into an explosive, this would certainly "assist" an aspiring nonnuclear weapons state in making such an explosive. No reasonable interpretation of the Nonproliferation Treaty would say that the Treaty intends, in exchange for an explicitly revocable promise by countries without nuclear explosives not to make them or acquire them, to transfer to them material that is within days or hours of being ready for incorporation into a bomb . . . . The NPT is, after all, a treaty against proliferation, not for nuclear development.

See Wohlstetter *et al.*, *Towards a New Consensus on Nuclear Technology*, 1979, pp. 34-35.

94. Christopher Ford, "NPT Article IV: Peaceful Uses of Nuclear Energy," Statement of the Principal Deputy Assistant Secretary of State for Verification, Compliance and Implementation, to the 2005 Review Conference of the Treaty on the Nonproliferation of Nuclear Weapons, New York, May 18, 2005, available from *www.state.gov/t/vci/rls/rm/46604.htm*.

95. *Ibid.*

96. According to Henry D. Sokolski,

> [The Principal Deputy Assistant Secretary of State for Verification, Compliance and Implementation] had attempted, to my knowledge, to get more cleared than what he was able to say [at the 2005 NPT Review Conference]. What he was able to say and what was cleared was that the United States at least is under no duty or obligation under Article IV to supply enrichment and reprocessing technologies to anyone. I think what he wanted to say might have included that countries really don't have a per se right to acquire this from others or to develop it even indigenously, but that was not approved.

See "Statement of Henry D. Sokolski," in "Assessing 'Rights'

Under the Nuclear Nonproliferation Treaty," a hearing before the Subcommittee on International Terrorism and Nonproliferation, Committee on International Relations, U.S. House of Representatives, 109th Congress, Second Session, Serial No. 109-148, March 2, 2006, Washington, DC: U.S. Government Printing Office, 2006, p. 28, available from *www.foreignaffairs.house.gov/archives/109/26333.pdf.*

97. See "Statement by the Office of the Director of National Intelligence on the North Korea Nuclear Test," Office of the Director of National Intelligence (ODNI) News Release No. 19-06, October 16, 2006.

98. North Korea acceded to the NPT on December 12, 1985, but officially withdrew from the treaty and terminated its IAEA safeguards agreement on January 11, 2002. See United Nations News Center, "Security Council Notified of DPR of Korea's Withdrawal from Nuclear Arms Accord," January 10, 2003; and IAEA, *Report of the Director General on the Implementation of the Resolution Adopted by the Board on 6 January 2003 and of the Agreement Between the IAEA and the Democratic People's Republic of Korea for the Application of Safeguards in Connection with the Treaty on the Non-Proliferation of Nuclear Weapons*, GOV/2003/4, January 22, 2003.

99. In February 1992, for example, the IAEA Board of Governors affirmed that the Agency's comprehensive safeguards agreements apply not only to declared nuclear material, but also to undeclared nuclear material that falls under the safeguards agreements and therefore should have been declared. See IAEA, *The Safeguards System of the International Atomic Energy Agency*, sec. B, para. 13, available from *www.iaea.org/OurWork/SV/Safeguards/safeg_system.pdf*. In May 1997, moreover, the IAEA Board approved the so-called *Additional Protocol*, a voluntary agreement which provides the Agency greater authority to inspect both the nuclear materials and the nuclear activities and technologies of states that have signed and ratified it. See IAEA, *Model Protocol Additional to the Agreement(s) Between State(s) and the International Atomic Energy Agency for the Application of Safeguards*, INFCIRC/540, Corrected, September 1997.

100. See Nikolai Khlebnikov, David Parlse, and Julian Whichello, "Novel Technologies for the Detection of Undeclared Nuclear Activities," IAEA-CN-148/32, released by NPEC in March 2007, available from *www.npec-web.org/Essays/20070301-IAEA-NovelTechnologiesProject.pdf*. Co-author Dr. Whichello serves as Head of the IAEA's Novel Technologies Unit of the Division of Technical Support's Section for Technical Support Coordination.

101. Henry D. Sokolski made this point when he testified before the House Subcommittee on National Security, Emerging Threats, and International Relations:

> The IAEA may be able to monitor nuclear fuel-making in rough terms, but it cannot inspect these facilities to provide timely warning of diversions or thefts equivalent to many nuclear weapons. It should admit this publicly. This would help put a spotlight on the dangers associated with additional governments trying to create even more nuclear fuel-making plants than already exist.

See Sokolski, "Clarifying and Enforcing the Nuclear Rules." Eldon V. C. Greenberg made a comparable point in 1993, when he wrote: "The NPT negotiators contemplated that safeguards had more than a merely formal role to play. If that role cannot be effectively fulfilled with respect to particular assistance or activities, then perhaps, in the presence of other factors, i.e., risk, lack of economic viability, such forms of assistance or activities should fall within the Treaty's prohibitions." See Greenberg, "The NPT and Plutonium," p. 19.

102. In November 2006, the IAEA decided to deny Iran (a government that is still in non-compliance not only with its NPT and IAEA safeguards obligations, but also several UN Security Council resolutions) technical cooperation at the latter's heavy-water reactor at Arak. This decision is a welcome development in the direction of increased candor from the Agency. For further background, see IAEA, *Cooperation Between the Islamic Republic of Iran and the Agency in the Light of United Nations Security Council Resolution 1737, 2006*, Report by the Director General, GOV/2007/7, February 9, 2007, available from *www.iaea.org/Publications/Documents/Board/2007/gov2005-7.pdf*.

103. French Republic, *Strengthening the Nuclear Non-Proliferation Regime*, Working Paper Submitted to the Preparatory Committee for the 2005 Review Conference of the Parties of the Treaty on the Non-Proliferation of Nuclear Weapons, NPT/CONF.2005/PC.III/WP.22, May 4, 2004, p. 2, available from *disarmament2.un.org/wmd/npt/2005/PC3-listofdocs.html*.

104. Christopher Ford, "The NPT Review Process and the Future of the Nuclear Nonproliferation Regime," Remarks by the U.S. Special Representative for Nuclear Nonproliferation to the NPT-Japan Seminar, "NPT on Trial: "How Should We Respond to the Challenges of Maintaining and Strengthening the Treaty Regime?" Vienna, Austria, February 6, 2007, available from *www.state.gov/t/isn/rls/rm/80156.htm*.

PART IV:

THE AGENCY'S AUTHORITY

CHAPTER 9

THE NUCLEAR NON-PROLIFERATION REGIME: AVOIDING THE VOID

Pierre Goldschmidt

Alexis de Tocqueville (1805-59) stated: "In politics what is often most difficult to understand and appraise is what is taking place under our eyes." De Tocqueville's insight suggests that it would be wise for the international community to stand back and to reflect on the lessons that should be learned from the International Atomic Energy Agency's (IAEA) experience in implementing safeguards over the last decade, particularly in North Korea and Iran. Such review and reflection will readily suggest that, ironically, just when the safeguards are getting better, the political will to use them effectively seems to be waning. Unless the IAEA is given the authority and tools to implement safeguards effectively and soon, the future of a rules-based approach for managing nuclear technology will dwindle and the prospects for sharing more widely the benefits of peaceful nuclear energy with developing countries may drop dramatically.

This chapter will explore how safeguards have gotten better, what lessons can be gleaned from the IAEA's experience over the last decade, and what solutions to the problems presented can be implemented by the international community.

I. SAFEGUARDS ARE GETTING BETTER

The IAEA safeguards system is being implemented more effectively and efficiently than ever before.

Traditionally, the IAEA focused on accounting for nuclear materials in a state facility-by-facility. This work was done only at declared facilities and was largely an audit. Since 1998, however, the IAEA has developed a global analytical approach that asks not simply whether the declared numbers add up, but also, "What's going on in this state's nuclear program? Is everything really consistent?"

At the heart of this approach is the production and periodic update of state evaluation reports (SERs) and of a corresponding action plan. SERs combine the results of inspections in the field and environmental swipes with analysis of all relevant information from open sources, including satellite imagery. State evaluation reports analyze the history of all anomalies and inconsistencies recorded during previous inspections. They examine whether a state's research and development program is internally consistent, corresponds with stated purposes, and points to a commitment to use nuclear technology exclusively for peaceful purposes. The SERs analyze export and import notifications regarding relevant nuclear material and equipment, and other information available to the IAEA. Every SER also includes a section that examines the most likely diversion scenarios, on the assumption that the state under review intends to divert nuclear material for military purposes.

Parallel with these developments, the IAEA has replaced almost all analog video cameras with digital surveillance cameras. Implementation of remote monitoring has increased from 14 systems in 2000 to 86 multicamera systems in 2004, and this trend is continuing. Progress is also being made in using more advanced equipment such as ground penetration radar to improve the IAEA's ability to verify that highly complex nuclear facilities conform to their official design. The IAEA has also established a new research and development (R&D) project to explore, with the support of member states, the potential use of advanced technologies in detecting undeclared nuclear material and activities.

In addition, in response to the discovery in 2004 of an extensive covert supply network of sensitive nuclear technology that came to light as a result of Libya's disclosure of its clandestine nuclear weapons program, the IAEA Department of Safeguards has established a new unit focused on documenting, investigating, and analyzing nuclear trade activities worldwide, with the aim of uncovering the existence of undeclared nuclear activities.

This more rigorous and resourceful approach to safeguards has led one knowledgeable commentator (Richard Hooper, *IAEA Bulletin*, June 2003) to assert in 2003 that "changes in structure and practices of the Safeguards Department have been accompanied by a change in culture that is more of a revolution than evolution." This "radical departure from the past practice" has also been acknowledged in the U.S. Government Accountability Office report of October 2005 on nuclear nonproliferation.

To be sure, there are still problems inherent in ensuring that, in "bulk facilities," even small amounts of nuclear material—a few kilograms among tons—are not diverted without timely warning, but the trend in the capacity of the safeguards system is clearly positive.

Unfortunately, the international community has failed to strengthen the authority of the IAEA to exercise its improved capacity in precisely the situations where it is most necessary: when a state has been found to be **in non-compliance** with its safeguards undertakings.

II. THE CASE OF NORTH KOREA

A. Summary of the IAEA's Experience with North Korea.

Soon after North Korea, formally the Democratic People's Republic of Korea (DPRK), concluded a comprehensive safeguards agreement (CSA) with the IAEA in 1992, the IAEA found the country to be in non-compliance. In 1993 North Korea gave notice of its withdrawal from the Treaty on the Non-Proliferation of Nuclear Weapons (NPT) as permitted under Article X. Negotiations between the United States and North Korea concluded in October 1994 with an "Agreed Framework," which averted a looming military-security crisis by inducing North Korea to freeze activity of its graphite-moderated reactors and related fuel cycle facilities in exchange for a U.S. commitment to deliver two 1,000-megawatt light water reactors (LWRs) and, in the meantime, to supply annually 500,000 tons of oil to meet heating and industrial needs. As part of this deal, North Korea remained a party to the NPT and the IAEA maintained a permanent presence monitoring the agreed freeze on nuclear activities.

The Agreed Framework, however, contained two provisions that sowed the seeds of the present potentially dangerous stalemate. First, it contained a clause that was interpreted by North Korea as limiting the IAEA's inspection rights under the CSA until such time as a significant portion of the LWR project was completed. Only then would the IAEA be allowed to take all the steps deemed necessary to verify "the accuracy and completeness of the DPRK's initial report on all nuclear material in the DPRK." Such limitation was clearly inconsistent with the lessons learned in Iraq that demonstrated that the IAEA needed greater access rights than those under the CSA and not the fewer rights embodied in the Agreed Framework.

The second flaw of the Agreed Framework was that it allowed North Korea to retain in storage all of its spent fuel containing weapons-grade plutonium and to maintain a reprocessing facility in a state of readiness so that North Korea could restart operations at any time. Only after completion of the LWR project would these facilities have to be dismantled. The U.S. negotiators and others recognized this flaw but could not persuade North Korea to remove it.

Because of the limitations in its inspection rights, the IAEA was unable to confirm that North Korea's initial declaration under its CSA was correct and complete. Therefore, every year **for 10 years**, North Korea was declared by the IAEA Board of Governors to be **in non-compliance** with its safeguards agreement. However, no additional penalties were imposed by the international community as a result of these declarations of non-compliance.

In 2002, the United States claimed to have discovered evidence that North Korea was developing an undeclared uranium enrichment program and, as a consequence, suspended the delivery of fuel oil under the Agreed Framework. In retaliation, North Korea expelled the IAEA's inspectors at the end of 2002 and withdrew from the NPT in January 2003. North Korea then reprocessed 8,000 (or more) spent fuel assemblies, and in 2004 declared that it possessed nuclear weapons.

Still, there have been no tangible consequences for these actions by North Korea beyond the isolation the country already experienced. China delivered substitute fuel oil to North Korea and threatened to veto any resolution of the United Nations Security Council (UNSC) adverse to North Korea. The six parties' talks initiated in 2003 have so far been chaotic and unproductive. As a result, 3 years after the IAEA inspectors were expelled from North Korea, they are still not allowed to return, and North Korea most likely has nuclear weapons. Meanwhile, the international community has not decided whether, from a procedural and legal point of view, North Korea has withdrawn from the NPT. This may sound like the discussion among religious scholars in 1453 on the sex of the angels while the Byzantine Empire was falling apart around them, but in actuality, knowing whether North Korea has or has not withdrawn from the NPT is more than an academic question.

If North Korea's withdrawal is acknowledged, then the IAEA should implement a limited safeguards agreement (INFCIRC/252) signed in July 1977 to verify a five megawatt thermal (MWth) research reactor delivered by the Soviet Union. This safeguards agreement, unlike a CSA, does not terminate when a state withdraws from the NPT. What could be verified there would, of course, be very limited, but it would be a matter of principle with potentially important consequences. If implemented, it would maintain at least a formal channel of communication between the IAEA and North Korea.

B. The Lessons Learned.

The three main lessons learned from the experience with North Korea are:

1. If a state withdraws from the NPT, any comprehensive safeguards agreement automatically terminates, and all nuclear materials and facilities are

no longer under safeguards and can be used freely and legally for a nuclear weapons program.

2. The threat of any permanent member of UNSC to use its veto right can block (for political, circumstantial reasons) any resolution adverse to a state withdrawing from the NPT.

3. A "voluntary, not legally binding freeze" of nuclear facilities gives no long-term guarantee that a state will not use them in the future.

C. What Are the Remedies?

The right to withdraw from the NPT remains a sovereign right. However, in order to minimize the consequences of such a withdrawal, the UNSC should adopt a **generic** resolution stating that, as a matter of principle, if a state is found by the IAEA to be in **non-compliance** with its safeguards undertakings **and** withdraws from the NPT **before** the IAEA has concluded (1) that its declarations are correct and complete; **and** (2) that there are no undeclared nuclear materials and activities in that state; **such a withdrawal constitutes a threat to international peace and security under Article 39 of the Charter of the United Nations**.

This **generic** resolution should also decide under Chapter VII of the UN Charter that any materials and equipment made available to such a state, or resulting from the assistance provided to that state, under a comprehensive safeguards agreement (INFCIRC/153-Corrected), will be removed from that state under IAEA supervision within 60 days of any notice of withdrawal from the NPT given by that State under Article X.1 of the NPT, and will remain under IAEA safeguards.

A withdrawing state should not be entitled to the benefits acquired while it was a party to the NPT and subject to comprehensive safeguards. This principle is not new. It is already contained in the IAEA Statute adopted in 1957, 13 years before the NPT came into force. Article XII.A.7 of the Statute states that "With respect to any Agency project, or other arrangement where the Agency is requested by the parties concerned to apply Safeguards, the Agency shall have the right . . . in the event of non-compliance **and** failure by the recipient State . . . to take requested corrective steps **within a reasonable time**, to suspend or terminate assistance and **withdraw** any materials and equipment made available by the Agency **or a member** in furtherance of the project" (emphasis added).

All nuclear-supplier states should also, in their bilateral nuclear supply agreements, reserve the right to require the return of all nuclear material and equipment previously supplied, in the event the recipient state withdraws from the NPT. One should bear in mind that withdrawing from the NPT is an option that Iraq has never threatened to use but that has been considered at the highest level of the Iranian leadership. The international community should not wait for the next crisis to happen before taking the appropriate preventive measures.

III. THE CASE OF IRAN

A. The Lessons Learned.

Without attempting to summarize here the findings of the IAEA with regard to Iran's previously undeclared nuclear activities contained in nine Reports to the Board of Governors and additional statements by the IAEA Deputy Director General for Safeguards (DDG-SG), suffice it to say that from these reports as well as eight Board Resolutions, one can draw three lessons: the need to avoid delaying tactics, the need to look beyond nuclear material, and the need to enforce transparency, each of which are more fully elaborated below.

1. The need to avoid delaying tactics.

In November 2003, Iran was found to be "in breach of its obligation to comply with the provisions of its safeguards agreement." This is synonymous to "non-compliance" and should have been reported to the UNSC as foreseen in Article XII.C. of the IAEA Statute. It was not reported mainly for two reasons. On the one hand, because of the fear of many member states that if the issue got out of the IAEA's hands and was reported to the Security Council, it would inevitably mean sanctions against Iran and that sanctions would lead nowhere except to another Iraq-like crisis which might well be a worse one. On the other hand, some member states feared that Russia and China could veto any resolution of the UNSC adverse to Iran. The worse would be for the IAEA to report Iran to UNSC only to have the issue blocked there, as was the case for North Korea, with no concrete outcome.

In October 2003, one month before the meeting of the IAEA Board of Governors, in order not to be referred to the Security Council, Iran agreed in Tehran with the EU-3 (France, Germany and the United Kingdom) to sign the Additional Protocol,[1] to implement it pending its ratification, and "to suspend all uranium enrichment and reprocessing activities as defined by the IAEA." However, less than 7 months later, on June 18, 2004, the Board of Governors adopted a resolution in which it deplored the fact that "as indicated by the Director

General's written and oral reports, Iran's cooperation had not been as full, timely and proactive as it should have been." And on September 18, 2004, the Board of Governors deeply regretted "that the implementation of Iranian voluntary decisions to suspend enrichment-related and reprocessing activities . . . fell significantly short of the Agency's understanding of those commitments and also that Iran has since reversed some of those decisions."

In November 2004, once more to avoid being reported to the UNSC and to gain time, Iran signed an agreement with the EU-3 in Paris, by which it decided, on a voluntary, not legally binding basis, to extend its suspension "to include all enrichment related and reprocessing activities" and "all tests or production at any uranium conversion installation." It was further stated that "the suspension will be sustained while negotiations proceed on a mutually acceptable agreement on long-term arrangements." However, on August 1, 2005, one day before receiving the EU-3 proposal, Iran announced its decision to resume uranium conversion activities.

So what is the situation today? Three years after the IAEA February 2003 visit to Natanz and the discovery of Iran's extensive undeclared nuclear program, there are still a number of outstanding questions due in large part to Iran's delaying tactics in providing access to locations, individuals and documents. As a result of these delaying tactics since the discovery of the Arak and Natanz sites in August 2002 and notwithstanding the Tehran and Paris suspension agreements with the EU-3:

- Iran has completed its conversion facility at Esfahan and produced a large quantity of uranium hexafluoride (UF6).

- Iran has introduced UF6 in the pilot enrichment plant at Natanz in June 2003, installed a 164-machine cascade by October 2003, manufactured more centrifuge components (1,274 assembled rotors at Natanz by October 2004), carried out work for the installation of the large underground enrichment facility at Natanz, and recently announced that it was resuming R&D-related enrichment activities.
- Iran is pursuing at full speed the construction of its heavy water research reactor, ignoring repeated requests by the IAEA Board of Governors to suspend it. This is of particular concern because the spent fuel of such a reactor will contain weapons grade plutonium.

On July 31, 2005, Dr. Hassan Rowhani, at the time Secretary of the Supreme National Security Council, presented his "performance report" to outgoing President Khatami. Referring to the Paris agreement of November 2004, he stated: "Since Iran had at that juncture completed its structural capabilities in the fuel cycle sector, it was possible to suspend the enrichment for a period of several months without making any fundamental damages to the fuel production project." There could be no clearer admission of the on and off strategy being followed by Iran.

2. The need to look beyond nuclear material.

The Director General's November 2004 report stated: "It should be noted that the focus of Agency Safeguards Agreements and Additional Protocols is nuclear material, and that, absent some nexus to nuclear material, the Agency's legal authority to pursue the verification of possible nuclear weapons related activity is limited." The limitation of the IAEA's focus on nuclear material is a major issue that has not been properly addressed by the international community. Much more than nuclear material is needed to build a nuclear weapon. Nuclear weaponization activities not involving nuclear material can be numerous and detectable.

Under a narrow legal interpretation of the IAEA's mandate and authority expressed by the language quoted above, effectively requiring **proof** that undeclared nuclear material and activities are related to a nuclear weapons program, the IAEA would have to find at least traces of nuclear material at an undeclared facility that can clearly be linked to equipment, material, or activities that could **only** be relevant to manufacturing nuclear weapons or other explosive devices. Such a narrow interpretation establishes a sleuthing standard that IAEA inspectors could hardly ever meet, and if such an interpretation prevails, the international community will be made ever more vulnerable to proliferation. A broader interpretation, certainly justified under the Agency's mandate to verify that nuclear material is not diverted to nuclear weapons or other nuclear explosive devices, which sees the Agency as having the authority to look beyond nuclear material itself, is the only interpretation under which the Agency can fulfil its mandate effectively.

Consider the limitations under the narrow interpretation. The sensitive equipment, material, and activities involved in a nonexclusively peaceful nuclear program would most likely be located at secret military sites. Yet, it is difficult, if not impossible, for the IAEA to access such sites **in a timely manner** under the standard CSA and even the Additional Protocol.

Experience has demonstrated that so many limitations can be imposed on IAEA inspectors when they get to such sites, that it is extremely unlikely that they would be able to **prove** that nuclear materials have been diverted to the manufacture of a nuclear explosive device. Even if such a conclusion could be drawn, it would likely be so late in the process of manufacturing nuclear weapons that it would be too late to deter the state from withdrawing from the NPT.

It is therefore essential for the IAEA to be understood to have the mandate and the authority to look for **any indication** that a non-nuclear-weapon state may be undertaking activities that could signal the existence of a nuclear weapons program, and to report such findings to the IAEA Board of Governors. It is encouraging to note that the IAEA Secretariat is progressively heading in that direction.

If a state intends to develop a nuclear-weapons capability it will need:

- to produce or acquire highly enriched uranium and/or weapons grade plutonium,
- to master all the necessary weaponization techniques, and
- to manufacture or acquire the required means of delivery.

There are indications that Iran is progressing on all three fronts. The following is known about Iran's weaponization activities and delivery-means, which go beyond its nuclear fuel cycle activities.

With respect to weaponization activities, the Director General's report to the IAEA Board of Governors dated November 18, 2005 (GOV/2005/87) indicates that among the documents received by Iran from intermediaries in 1987 was one related to "the casting and machining of enriched . . . uranium metal into hemispherical forms." Such a process has no peaceful application and therefore represents a substantial indication that Iran has been (and may still be) interested in developing a nuclear weapons capability. The DDG-SG also reported on January 31, 2006, that the Agency had information about tests related to high explosives that could have a military nuclear dimension. In addition, efforts by the Physics Research Center (an organization related to the Iranian Ministry of Defense that was located until 1998 at the now razed Lavizan-Shian site) to acquire dual use materials and equipment that could be used in uranium enrichment and conversion activities, is another relevant indication. Interestingly, a commentary published on February 12, 2006, in the conservative Iranian daily *Keyhan* argues that "benefiting from the knowledge of and ability to manufacture nuclear weapons is something different from the triple issues of producing, storing, and using such weapons. However if necessary . . . then the ground will be paved for moving toward the subsequent phases."

With respect to delivery means, it should be noted that aside from the five nuclear weapons states and the three non-NPT states, only three countries: North Korea, Saudi Arabia, and Iran, are known to possess medium to long-range ballistic missiles capable of carrying a payload of 1,000kg or more, sufficient for a nuclear warhead. In his briefing dated January 31, 2006, to the IAEA Board of Governors, the DDG-SG indicated that Iran rejected a request to discuss information available to the Agency about "the design of a missile re-entry vehicle . . . which could have a military nuclear dimension."

3. The need to enforce transparency.

Not only must the IAEA's evidentiary lens be widened, the transparency measures for which it calls must be made enforceable. The Director General in his report of September 2, 2005 to the IAEA Board of Governors states:

> In view of the fact that the Agency is not yet in a position to clarify some important outstanding issues after two and a half years of intensive inspections and investigation, **Iran's full transparency is indispensable and overdue**. Given Iran's past concealment efforts over many years, such transparency measures should extend beyond the formal requirements of the Safeguards Agreement and Additional Protocol and include access to individuals, documentation related to procurement, dual use equipment, certain military owned workshops and research and development locations. **Without such transparency measures**, **the Agency's ability** to reconstruct, in particular, the chronology of enrichment research and development, which is essential for the Agency to verify the correctness and completeness of the statements made by Iran, **will be restricted** (emphasis added).

Since 2003 the IAEA Board of Governors has adopted a half dozen resolutions calling on Iran to be more transparent and cooperative. In its last resolution of September 24, 2005, the IAEA Board of Governors "urges Iran to implement transparency measures, as requested by the Director General in his report."

Unfortunately, such requests by the Board of Governors have no legal force and effect and do not allow IAEA inspectors to obtain broader access to individuals, documents, or locations.

B. What Are the Remedies?

The single most effective and feasible way to establish the necessary measures is for the UNSC to adopt a **generic** and binding resolution stating that if the IAEA finds a State in non-compliance **and** requests increased verification authority, the UNSC would automatically adopt a **specific** resolution (under Chapter VII of the UN Charter) providing this additional authority until the IAEA has concluded that there is no undeclared nuclear material and activity in that State and that its declarations are correct and complete. If such a **generic** resolution existed in November 2003, it may well be that the IAEA Board of Governors would not have been afraid to declare Iran in non-compliance and would have reported Iran to the UNSC **for the sole purpose** of requesting such broader verification authority, which clearly has nothing to do with sanctions.

Such a **generic** resolution should also request the non-compliant state to **suspend** all sensitive nuclear fuel cycle activities at least until the Agency has been able to draw the above mentioned conclusion, or, possibly, for automatically renewable periods of 10 years unless otherwise decided by the UNSC. This would be what Dr. ElBaradei has called a "rehabilitation period" or a "probation period, to build confidence again, before you can exercise your full rights." (cf. interview with *Newsweek*- January 23, 2006)

Independently, the Nuclear Supplier Group could adopt a rule whereby nuclear material and equipment would only be exported if the facilities where they are to be stored or used are covered by both a comprehensive safeguards agreement and an INFCIRC/66-type safeguards agreement. This requirement would block a recipient state from withdrawing from the NPT and claiming the right to do whatever it wants with the items previously delivered or the materials derived therefrom.[2]

IV. CONCLUSION

The IAEA Statement at Main Committee II of the NPT Review Conference in May 2005, states:

> As underlined by the Director General in his opening statement, our verification efforts must be backed by an effective mechanism for dealing with cases of **non-compliance** with Safeguards Agreement or of **withdrawal** from the NPT. For this, both the NPT and the IAEA Statute make clear our reliance on the Security Council to promptly consider the implications of such cases for international peace and security and to take appropriate measures.

As suggested in this chapter, concrete measures can readily be taken within the IAEA and UN framework to improve the assurance that all nuclear material and activities in a non-nuclear-weapon-state **found to be in non-compliance** are and remain exclusively for peaceful purposes. The UNSC can take these vital generic measures without eroding state sovereignty or development. The measures proposed here would apply **only** when the highly representative IAEA has found a state to be in **non-compliance** with its safeguards obligations. None of these measures would impede a state's right or capacity to enjoy the peaceful uses of atomic energy. On the contrary, these measures would quicken the international community's capacity to regain confidence that a state that may have wandered off the peaceful nuclear path had corrected its course and would once again be a reliable neighbour and business partner.

Without UNSC action of this sort, the future of a rules-based approach for managing nuclear technology will dwindle, and the prospects for sharing the benefits of peaceful nuclear energy more widely with developing countries may drop dramatically. Inaction is playing against the credibility of the NPT regime.

As Cardinal de Richelieu once said: "Politics is the art of making possible what is necessary."

ENDNOTES - CHAPTER 9

1. The Model Additional Protocol, INFCIRC/540 (corrected) approved by the IAEA in March 1997 provides for increased reporting by a state on its nuclear fuel cycle related capabilities and activities and expanded short notice IAEA access to nuclear-related locations.

2. A CSA remains in force only for so long as the state remains party to the NPT, whereas under a INFCIRC/66-type agreement, all nuclear material supplied or produced under that agreement would remain under safeguards even if the state withdraws from the NPT until such time the IAEA has determined that such material is no longer subject to safeguards.

CHAPTER 10

THE U.S.-INDIA CIVIL NUCLEAR COOPERATION INITIATIVE: THE QUESTION OF SAFEGUARDS

Quentin Michel

Since the public of announcement in 2004 by Indian Prime Minister Manmohan Singh and U.S. President George W. Bush of a civil nuclear cooperation initiative, the question of potential cooperation with India has been heavily debated. It has acted as an incentive for other major supplier states[1]—essentially the nuclear weapons holders—to conclude or announce their intention to complete similar agreements.[2]

Voices from others suppliers have also declared that such cooperation will breach most of the international commitments they have contracted. In both cases, we have to admit that the U.S.-India agreement is not just a bilateral question but has become a multilateral one, and its impact on international export control regimes will have a major significance. This discussion will not be devoted to the content of the agreement itself but more on how an exception for India could be possible without breaching international export control regime(s). Furthermore, we will focus only on the conditions of supply and in particular on International Atomic Energy Agency (IAEA) safeguards as required by international export control regimes—mostly the Nuclear Non-Proliferation Treaty (NPT) and the Nuclear Suppliers Group (NSG). It should be noted that other safeguards mechanisms could be imposed on the recipient state. We could mention safeguards required by bilateral safeguards agreements between suppliers and end users similar to the one offered by U.S. authorities in the early 1950s in the implementation of the Atoms for Peace plan. Secondly, there is the possibility that the IAEA could assume safeguards implementation on behalf of a bilateral agreement as defined by Article 12 of IAEA statutes. Finally, we have several potential bilateral fallback agreements in case of breach of the initial IAEA safeguards agreement.[3]

NPT Conditions of Supply for Nuclear Items.

The conditions of supply of nuclear items to non-NPT states are established by Article III.2 which states that: "Each State Party to the Treaty undertakes **not to provide**: (a) source or special fissionable material, or (b) equipment or material especially designed or prepared for the processing, use or production of special fissionable material, to any non-nuclear-weapon State for peaceful purposes, **unless** the source or special fissionable material **shall be subject to the safeguards required by this article**."

Since the entry into force of the NPT, states have always argued on the category of safeguards to request of the recipient state before transferring the items. If it was clear from the reading of Article III.1 that it should be organised within the framework of the IAEA safeguard system, the content, and in particular its field of implementation, was not clearly defined, so for a majority of nuclear supplier states, safeguards requirements should only apply to transferred nuclear material and to nuclear material used by the transferred nuclear items (equipment or technology). Such safeguard requirements were in the line of the guidelines of the Zangger Committee, which was considered by a majority of NPT states as their informal interpretation body.[4]

Nevertheless, this approach was rather anachronistic considering that safeguards required from non-NPT state parties were less comprehensive than the one required from states that were parties to the NPT. The NPT ratification required that a non-nuclear-weapons state conclude with the IAEA a full scope safeguards agreement, which would apply to all its nuclear material use in all its peaceful activities and not only to the nuclear items transferred.[5] This safeguards system is also known as the Comprehensive Safeguards Agreement (CSA).

This discriminatory approach has had a counterproductive effect due to the fact that it granted indirectly a privileged treatment to non-NPT states by conceding them less severe verification requirements.

So in 1995, the NPT Review Conference has reviewed this interpretation of Article III.2 to align it to safeguards applied to NPT non-nuclear-weapon states. In the document *Principles and objectives for nuclear non-proliferation and disarmament*, it is affirmed that

> New supply arrangements for the transfer of source or special fissionable material or equipment or equipment or material especially designed or prepared for the processing, use or production of special fissionable material to non-nuclear-weapon States should require, as a necessary precondition, acceptance of the Agency's full-scope safeguards and internationally legally binding commitments not to acquire nuclear weapons or other nuclear explosive devices.[6]

Therefore, with this interpretation of Article III.2, transfers of nuclear material and equipment to a non-NPT state party like India will be ruled by a CSA agreement into force before the transfer could take place. So if before 1995, transfers to India could have occurred with a dedicated safeguards mechanism defined by INFCIRC/66, presently a transfer to India could only be possible if India concludes a CSA agreement with the IAEA. Considering the U.S.-India Civil Nuclear Cooperation Initiative, a CSA agreement as defined by the INFCIR/153 could not be possible due to the fact that Indian nuclear activities submitted to IAEA safeguards will be defined in a list of civilian facilities established by the Indian authorities. Such a mechanism appears to be similar to the voluntary safeguards agreement taken by an NPT nuclear weapons state where the list of peaceful facilities submitted to IAEA safeguards is provided by the state.[7] The objective to submit nuclear-weapons states to a verification mechanism is more to avoid the risk of unfair competition and balancing the administrative and commercial burden that non-nuclear-weapons states have to face rather than to control the risk of diversion of peaceful nuclear facilities. Therefore, such a voluntary mechanism could not be applied to India unless it were to consist of informally granting the status of an NPT nuclear weapons state. In this regard, the provision of Article IX of the NPT leaves no room for interpretation: only states which have manufactured and detonated a nuclear weapon prior to January 1, 1967, can hold such a status.

To conclude, the implementation of the U.S.-India Civil Nuclear Cooperation Initiative will not be possible without breaching safeguards requirements established by Article III.2 of the NPT as highlighted by NPT review conferences and the Zangger Committee.

Nuclear Suppliers Group Conditions of Supply for Nuclear Items.

The most important informal instrument regarding the control of nuclear trade is the Nuclear Suppliers

Group (NSG).[8] Contrary to the Zangger Committee, the NSG is not informally linked to the NPT. The NSG does not establish an international nuclear export control regime, its main objective is in the definition of a common understanding of export control principles that each participating state will introduce in its national export control regime.

The NSG has adopted two groups of guidelines. The first set of guidelines (the trigger list)[9] governs the export of items that are especially designed or prepared for nuclear use, and the second governs the export of nuclear-related dual-use items and technologies, that is, items that can make a major contribution to an unsafeguarded nuclear fuel cycle or nuclear explosive activity, but which also have non-nuclear uses, in the chemical industry for instance.[10] Concerning potential transfers to India as defined by the U.S.-India Civil Nuclear Cooperation Initiative, only the NSG guidelines governing the transfer of nuclear items (the trigger list) will apply.

In conformity with paragraph 4 of the NSG trigger list guidelines, the supplier state should, before granting the export authorization, verify if the state end-user fulfils the different export conditions defined by the NSG guidelines. One of the main conditions of supply concerns the obligation of the end-user to have brought into force a CSA agreement with the IAEA requiring the application of safeguards on all sources and special fissionable material in its current and future peaceful activities.[11] It should also be noted that if the NSG considers "that the provisions of the IAEA model Additional Protocol[12] will strengthen the nuclear safeguards regime and facilitate the exchange of nuclear and nuclear related material in peaceful nuclear cooperation,"[13] it does not require it yet as a condition of supply. Although this question has been analyzed systematically by the subsequent plenary meetings, no consensus has been obtained between participating states. To resolve this ever-lasting discussion, an approach to bring about the potential entry into force of such a condition of supply has been proposed to the participating states, but the necessary consensus has not yet been reached.

The NSG trigger list Guidelines establish two exceptions to its CSA requirement for transfers of nuclear items to a non-nuclear-weapon state. The first is a classical "Grandfather" clause,[14] which authorizes NSG supplier states not to require a CSA to agreements or contracts drawn up before their date of adherence. The second is the so-called "safety clause," which authorizes NSG supplier states to transfer nuclear trigger list items to a non-nuclear-weapon state only in exceptional cases and if they are deemed essential for the safe operation of existing facilities and if dedicated safeguards are applied to those facilities. Moreover, before granting such authorization, suppliers should inform and, if appropriate, consult with the other NSG participating states in the event that they intend to authorize or to deny such transfers.

This exception has been used only twice by Russia to supply fissile material for a nuclear power plant to India in 2000 and 2006. For the first Russian fuel shipment to India, most of NSG members states expressed concern that such an exception could only be used when the assistance by an NSG member state is essential to prevent or correct an imminent radiological hazard that poses a significant danger to public health and safety. Such conditions were, for them, obviously not met in the export of Russian fuel to India. Therefore, a process was initiated to strengthen and obtain a commonly-agreed interpretation of the safety clause and in particular on the terms "exceptional cases." However, the NSG did not succeed in adopting a common interpretation. In 2006, when Russia announced its intention to again use the safety clause to export nuclear fuel to India, NSG member states appeared less concerned by the transfer. This rather consensual reaction could only be explained by the new NSG-India relationship initiated by the U.S.-India Civil Nuclear Cooperation Initiative and other similar declarations made by other nuclear weapons states.

Nevertheless, paragraph 4 of the NSG trigger list guidelines and, therefore the CSA condition, does not apply to transfer of nuclear items to nuclear-weapon states. The guidelines did not contain specific provisions on the category of guidelines to be required by the supplier when it intends to export trigger items to a nuclear-weapon state. Consequently, it is up to the supplier state to define the safeguards requirements it intends to impose on the recipient. For transfers to NPT nuclear weapons states, the situation is rather simple so long as all of them have signed a voluntary safeguards agreement with the IAEA including specific provisions implementing the additional protocol.

Considering that NSG Guidelines for Nuclear Transfers contain no reference to the NPT[15] and, therefore, no reference to the NPT definition of a nuclear-weapons state, how does the NSG define a nuclear-weapons state? In other words, could it be possible that the NSG definition of a nuclear-weapons state will be broader than that of the NPT?

The absence of any reference to the NPT in the NSG guidelines is mostly due to historical reasons. In 1978, when the NSG was created, France was not a NPT member and set as a condition of its adherence to the

NSG that its guidelines contain no explicit reference to the NPT. Nevertheless, the lack of reference to the NPT has not been completed by a definition of non-nuclear and nuclear-weapons states in the guidelines. Moreover, nothing in the guidelines prohibited NSG participating states from adopting a definition of a nuclear-weapons state that could include India, Pakistan, or Israel. If we approve this assumption, what will be the safeguards required by the supplier to transfer nuclear items to a nuclear-weapons state? As in the case of the NPT nuclear-weapons states, the safeguards requirement will be defined on a national basis by the authorities of the supplier state. Nevertheless, if nuclear transfers to non-NPT nuclear-weapons states could in theory be envisaged, the current practice of the NSG does not work with such an interpretation. Most of the NSG participating states export authorization denials concern non-NPT nuclear-weapons states.

Finally, all NSG participating states are presently parties to the NPT, and nuclear transfers to non-NPT nuclear-weapons states like India could not be authorized, considering the different commitments they have taken with their NPT ratification.

Conclusion.

Considering the safeguards condition of supply of the two main formal and informal international nuclear export control instruments, we do not see how the U.S.-India Civil Nuclear Cooperation Initiative will be implemented by the supplier state without breaching their safeguards commitments. Even if the cooperation is submitted to the entry into force of an India-specific safeguards agreement negotiated with the IAEA that will control all civilian nuclear facilities in perpetuity, it does appear that such a specific agreement will never conform to the CSA required by both international export control regimes. The Indian commitment to adhere to and sign an additional protocol does not change the situation because, once more, it will concern only Indian civilian facilities as listed by India.

It should be recalled that if the NSG could in the medium term, by its absence of reference in its guidelines to the NPT, adopt an exception to allow nuclear transfer to India, its participating states will find it difficult to individually implement such exceptions due to their legally binding NPT commitment.

It remains to be seen if nuclear supplier states are ready to embark in this new nuclear nonproliferation approach initiated by the U.S.-India agreement based on the political cooperation strengthening between suppliers, even if it will induce the infringement of their NPT commitment.

ENDNOTES - CHAPTER 10

1. British Prime Minister Tony Blair warmly welcomes the U.S.-India nuclear deal, available from *www.number-10.gov.uk/output/Page9124.asp*. French President Jacques Chirac and Indian Prime Minister Singh made a common declaration on a potential nuclear peaceful agreement in February 2006, available from *www.diplomatie.gouv.fr/actu/bulletin.asp?liste=20060220.html#Chapitre9*; in November 2006, China and India signed a civilian nuclear cooperation deal. In January 2007, Russian President Vladimir Putin signed an agreement with New Delhi to reinforce Russia's nuclear peaceful cooperation with India, available from *www.washingtonpost.com/wp-dyn/content/article/2007/01/25/AR2007012500182.html*.

2. Australian Prime Minister John Howard has recently joined the group by expressing Australia's willingness to sell uranium to India provided New Delhi adheres to strict safeguards, available from *www.khaleejtimes.com/DisplayArticleNew.*

asp?xfile=data/subcontinent/2007/April/subcontinent_April100.xml§ion=subcontinent&col=.

3. See for example the new formulation of Articles 4 (a) and 16 of the INFCIRC/254.Rev.8 Part.1, available from *www.iaea.org*.

4. At the first Review Conference of the NPT in 1975, a brief paragraph in the final document mentioned the work of the Zangger Committee by referencing the IAEA document publishing its guidelines. This paragraph stated:

> With regard to the implementation of article III, paragraph 2 of the Treaty, the Conference notes that a number of States suppliers of material or equipment have adopted certain minimum, standard requirements for IAEA safeguards in connection with their exports of certain such items to non-nuclear-weapon States not party to the Treaty (IAEA document INFCIRC/209/Rev.2). The Conference attaches particular importance to the condition established by those States, of an undertaking of non-diversion to nuclear weapons or other nuclear explosive devices, as included in the said requirements.

5. A model of such safeguards agreement has been established by IAEA under the reference INFCIRC/153.

6. Paragraph 12 of Decision 2 (NPT/CONF.1995/32(Part I) Annex).

7. See for the United States, United Kingdom, France, China, and Russia, respectively, INFCIRC 288, INFCIRC 263, INFCIRC 290, INFCIRC 369, and INFCIRC327.

8. Participating NSG States are Argentina, Australia, Austria, Belarus, Belgium, Brazil, Bulgaria, Canada, China, Cyprus, Czech Republic, Denmark, Estonia, Finland, France, Germany, Greece, Hungary, Ireland, Italy, Japan, the Republic of Korea, Latvia, Lithuania, Luxembourg, Malta, Netherlands, New Zealand, Norway, Poland, Portugal, Romania, the Russian Federation, Slovakia, Slovenia, South Africa, Spain, Sweden, Switzerland, Turkey, Ukraine, the United Kingdom, and the United States.

The European Commission participates as an observer. Available from *www.nsg-online.org/*.

9. The list contains the following categories: nuclear material; nuclear reactors and equipment therefore; non-nuclear material for reactors; plant and equipment for the reprocessing, enrichment, and conversion of nuclear material and for fuel fabrication and heavy water production; and technology associated with each of the above items. The guidelines have been published by the IAEA under the reference INFCIRC/254Part.1.

10. The list of concerned items has been divided into six categories: industrial equipment, materials, uranium isotope separation equipment and components, heavy water production plant related equipment, test and measurement equipment for the development of nuclear explosive devices, and components for nuclear explosive devices. The guidelines have been published by the IAEA under the reference INFCIRC/254Part.2.

11. Paragraph 4(a).

12. See INFCIRC/540.

13. See Press Statement of NSG Plenary Meeting, Paris, June 22-23, 2000, available from *www.nsg-online.org/PRESS/2000-Press.pdf*.

14. INFCIRC/254/Rev.6/Part 1, Paragraph 4(c).

15. The second group of guidelines dedicated to the export of nuclear dual-use items mentions only twice the NPT in paragraph 4, which is dedicated to criteria that supplier states have to consider in the decision process to grant or not grant the export authorization.

CHAPTER 11

FINANCING IAEA VERIFICATION OF THE NUCLEAR NONPROLIFERATION TREATY

Thomas E. Shea

Introduction.

Nations spend billions on defense, but the amount the international community spends to finance International Atomic Energy Agency (IAEA) verification of the Nuclear Nonproliferation Treaty (NPT) in all states is only $120M/year.[1] The provisions for financing IAEA programs are set out in the Statute of the Agency, and that arrangement has proven to provide adequate funds to sustain the program and to bring the effectiveness of the safeguards system to its current capabilities.

The IAEA enjoys enormous international prestige and is held up within the United Nations (UN) family as a model of efficient operation. Now that the Democratic People's Republic of Korea (DPRK) has carried out a nuclear test, is there an opportunity to reconsider whether the Agency should be asked to do more, and whether added investments in it would help to bolster the nonproliferation regime?

There may be a number of areas where the Agency might take on additional capabilities or improve its current performance if the Agency had additional money, and in some cases, additional authority. DPRK provides a clear justification for the types of activities mentioned, and I am optimistic that should the Director General ask for significant safeguards expansions and upgrades, the funding will be forthcoming. To my mind, the Director General should convene a council of wise men to assist in determining how best to respond in this matter.

In addition to the areas addressed earlier today, the Agency needs to replace its Safeguards Analytical Laboratory and wishes to accelerate the turnaround time for environmental samples. It needs to implement advanced data visualization systems to analyze and evaluate the streams of data arising from open source information analysis and other modern safeguards methods. It should also bolster the NPT regime by: (1) strengthening international norms against proliferation; (2) assuring the human capital needed to carry out the myriad tasks associated with implementing the nonproliferation regime; (3) facilitating or even stimulating the global expansion of nuclear power while providing compelling advantages to states to refrain from acquiring sensitive nuclear technologies; (4) developing and deploying nuclear power systems tailored to the needs and challenges of the developing areas of the world—where future problems are most likely to emerge; and (5) beginning constructive steps in relation to the disarmament commitments of the nuclear-weapons states parties to the NPT, and extending that enterprise to include all states possessing nuclear weapons.

These roles could have fundamental and significant impacts on international security; they would cost from tens of millions to billions of dollars or Euros per year to realize.

Financing IAEA Safeguards: Existing Practice.

Under paragraph 2 of INFCIRC/153, the Agency is *obligated* to ensure that safeguards will be applied in accordance with the terms of the safeguards agreements. Safeguards in non-nuclear-weapons states concluded pursuant to INFCIRC/153 must be applied; member states must pay the fees assessed under the provisions of the IAEA Statute[2] as part of the regular budget to ensure that the Agency is able to meet its obligations. All other IAEA programs are voluntary in nature and depend upon the availability of adequate resources to be carried out.

The existing financial system provides a reliable funding stream for the regular budget assessments once established; the challenges arise in:

- defining just what safeguards are actually necessary to meet these obligations, and,
- the difficulty in achieving increases in the regular budget when additional activities, staff or equipment are considered necessary.

During the long lean years, the Safeguards Department lived on zero real growth, coping by introducing technical innovations that improved verification coverage and quality—equipping inspectors and inspection systems with computers and getting facility operators to make their declarations on computer media that can be read by inspector-computers at the facilities during inspections. The Safeguards Department also gained efficiencies by deploying its inspectors increasingly through regional offices as a way to increase the days an inspector can actually spend inspecting, by reducing or cutting out inspection activities that are optional (such as in nuclear-weapons states) and by changing the safeguards rules and procedures to either reduce the requirements or to find alternative means to secure the assurances needed. Pierre Goldschmidt managed to secure a substantial increase in the regular safeguards budget, but it took years before the Board was finally convinced.

The regular budget for financing the IAEA is governed by the provisions set out in the Agency's Statute, which each member state accepts. Each year, as safeguards is a mandatory program, budget estimates of what it will cost to meet the required verification activities are prepared, based on guidance from the director general and a sense conveyed informally from the Geneva Group.[3] Sometimes the guidance comes first, sometimes it is a reaction representing what the director general senses the traffic will bear. Following internal consultations and adding the required shares to support the management activities and other costs, the director general presents the budget to the Program and Budget Committee of the Board in May of each year. When the Committee is satisfied, it recommends the budget to the Board, and when it is satisfied, the Board submits the recommended budget to the General Conference for its approval.

Budget increases are resisted for a host of reasons. National treasuries always have competing demands. In addition to resisting expenditures simply due to competing demands, IAEA member states are normally not seeking to expand the power of international organizations, as sooner or later the power and influence they achieve might be exercised against a state's national interests. Preventing mission creep remains an active concern. Also, achieving an increase in the regular IAEA safeguards budget also involves maintaining some sort of balance with contributions to technical cooperation. Moreover, when cuts in other programs have been proposed as a means to provide additional money for safeguards, the director general has refused.

There are ways to mobilize a consensus to demonstrate that additional capabilities are needed. The director general convenes wise-men meetings from time to time; there are internal and external audit requirements (financial and programmatic) to assure that the ship remains on course. The U.S. General Accounting Office carries out independent program reviews to determine for the U.S. Congress that its appropriations are providing the capabilities it seeks.

All things considered, the Agency's verification capabilities are today vastly superior to where they stood when the NPT came into force, or when Iraq and the DPRK first violated their nonproliferation undertakings.

Extrabudgetary Contributions.

In addition to the regular budget, the IAEA relies on extrabudgetary contributions from its member states. In 2005, member states provided extrabudgetary contributions in the amount of $130,863,115 to the Agency in cash and in kind.[4] Most of this is for the Technical Cooperation Fund, but some of it goes to the Safeguards Department—not for mandatory inspections, but for equipment or inspections in nuclear-weapons states, for example. The U.S. voluntary contribution to the IAEA in 2006 was $49.5 million; $19.1 million of that was for safeguards and $14.2 million was for the U.S. Program of Technical Assistance to Agency Safeguards (POTAS). Counting POTAS, there are about 18 member state support programs that provide money and talent for the Safeguards Department to improve its capabilities and performance.

Extrabudgetary contributions are also provided by other UN organizations and other international organizations, in the amount of $6.8M in 2005. This included a contribution by the Nuclear Threat Initiative (NTI).[5] At the special event that took place during the 2006 IAEA General Conference, former Senator Sam Nunn, Co-Chairman of NTI, announced a contribution of $50M to be used by the IAEA, together with other contributions, to establish a nuclear reactor fuel bank that would provide assurances of supply to states adopting nuclear power. NTI represents a new departure for the IAEA, a philanthropic institution investing in the IAEA to accomplish activities related to nuclear security and nonproliferation issues.

The Agency has a policy in place to accommodate contributions from virtually any source, assuring that the Agency's policymaking organs will determine how such funds are managed and spent.[6] Note that it is not common for the IAEA to solicit funds for activities that are not supported by existing mechanisms. However, the Board, acting on a request by the director general, did establish a special fund for the receipt of the Nobel Peace Prize, the "IAEA Nobel Cancer and Nutrition Fund." In establishing the fund, "the director general also encourages member states and other donors to contribute to the special fund by making available additional resources both in cash and in kind, to be used to maximize the Agency's ability to build capacity and transfer the needed know-how to developing countries."[7] Thus, a precedent—albeit limited—has been established in which the Agency has gone beyond the normal financial means available to it to encourage donations from unspecified parties.

Expanding the Nonproliferation Regime on a Different Financial Basis.

Increasing contributions from national treasuries could be significant if there is a proliferation event—such as the DPRK nuclear test—or if a new treaty comes into force that carries financial obligations with it. Short of that, further increases are likely to be sporadic, driven when a consensus eventually emerges demanding improvement.

However, there is another way. Suppose that the nonproliferation regime provided a steady stream of significant income so that the decision shifted from how to raise money to how to spend it. The whole notion of creative steps to strengthen the nonproliferation system would then appear in a different light.

Here are five ways in which such a condition could be created.

1. *Endowment:* A "Nonproliferation Endowment" could be chartered to improve the IAEA's ability to verify the NPT and to stimulate peaceful nuclear programs designed for economic development and a stable peace. Such an endowment could be funded by substantial donations from wealthy individuals or foundations. Such an effort would actively solicit contributions from the public, the nuclear industry, the alumni of the nonproliferation work force, and governments as well. Note that the Harvard endowment, which includes some 10,000 contributions, is now valued at approximately $26B.

2. *Surcharge:* In the United states, "customers who use nuclear power pay for the disposal of spent fuel. The federal government collects a fee of one mil (one-tenth of a cent) per kilowatt-hour of nuclear-generated

electricity from utilities. This money goes into the Nuclear Waste Fund. In addition, Congress makes an annual appropriation from the General Fund of the Treasury to pay for disposal of defense-related high-level radioactive waste."[8] Of course, any country could establish a surcharge system for any reason, like spent fuel management, or for nonproliferation purposes. Today, there are approximately 449 nuclear power plants in operation;[9] if that were the case when the IAEA was created, it is possible that the Statute might make different arrangements. The Agency's Statute could be revised, possibly to make a surcharge on all plants constructed after a specified date.

A surcharge arrangement might fit best into a new legal framework, as a basis for transparency-related measures under a fissile material cut-off treaty, for example, or under a future framework for expanding global nuclear power as a means to stimulate nuclear power in the developing areas of the world. In the latter case, such a funding stream might be used to start-up new nuclear projects under a scheme that allowed delayed repayment such that the nuclear plant could begin to bring about economic development for several years before repayments would commence.

A surcharge should be levied as a fixed percentage of some commodity price. That way, the rate is the same for all states or exporters, and the amounts of money would follow inflation in a natural way without the need for periodic negotiated adjustments with all the drama that such steps would entail. For example, a surcharge of 1 percent on nuclear generating costs collected from nuclear utilities would provide an funding stream of $700M/year from the United States alone.[10] One percent may be too much or too little; only by considering the aims for such a framework could a defensible figure be set.

3. *Selling services:* The IAEA could be asked to organize nuclear operations under extra-territorial agreements with host countries. These might include nuclear power plants for regional power sharing in the developing areas of the world, multinational uranium enrichment centers, spent fuel reception centers, multinational spent fuel recycle centers and nuclear waste repositories. The Agency's role in such cases would be to provide the political framework and to secure competent commercial organizations to actually operate the respective facilities. In such cases, it would be reasonable for the Agency to collect fees for the services it provides.

4. *Financial Institutions*: A financial institution (like the World Bank) could be empowered to engage in financing appropriate peaceful nuclear projects under a delayed payback arrangement. The World Bank itself does not currently finance nuclear projects; it did once in Italy,[11] and today the World Bank is carrying out an investigation to determine whether or not to re-enter this field.[12]

Whether the World Bank or one or more other financial institutions, such an arrangement would depend upon the capitalization provided and time-dependent returns. The delayed repayment scheme identified above would be appropriate, but in addition, consideration might be given to having the financial institutions actually purchase and own the power plants, transferring ownership upon repayment. Such an arrangement would ensure that vendors would receive payments, that prices would be fair, that users would have a measure of assurance of supply, and that vendors could be provided with some degree of indemnification against spurious litigation. Investments made by the financial institution might also carry an accompanying contribution to the IAEA to cover its expenses as necessary to ensure that the project serves the intended purposes and that the quality of goods and services provided is consistent with international standards.

5. *Market Mechanisms*: The fourth possibility would somehow engage the investment community through the issuance of tax-exempt nonproliferation bonds, which would yield interest on revenues collected down-stream by financing projects under the delayed pay-back arrangements described above. This scheme might connect with one of the earlier mechanisms and would require government investment and oversight to be stable and to avoid suspicions that it might be a ponzi scheme.

6. *Industry Share*: This proposal goes directly to provide the IAEA with enhanced technical capabilities by engaging the exporters of nuclear facilities. Under current practice, if a state imports a reactor or fuel manufacturing plant or any other type of fuel cycle facility, the importer is required to submit the facility for IAEA safeguards. The facility operator and the Agency bear costs as necessary for safeguards to be applied; sometimes the state bills the Agency for the installation of safeguards equipment, sometimes not. The facility operator may pass the costs along as business expenses to its customers.

Under such an arrangement, for plants to be exported, the vendor and the future facility operator would work with the Agency to develop a safeguards approach, including the inspection equipment to be used by the Agency and the procedures for its maintenance and operation. The vendor would then be responsible for providing such equipment that would become part of the sales price. To the extent that the vendor remains engaged for the maintenance or operation of any plant systems, the vendor would remain responsible for assuring that the safeguards equipment continues to meet IAEA needs, including maintenance and upgrades as appropriate. Just as for plant safety systems, the safeguards systems should be integrated into the plant operational systems such that continued operation would be prevented in the event of anomalous indications from the installed safeguards systems.

Conclusions.

The premise of my remarks has been that the proliferation of nuclear weapons threatens national and international security, and that, as the renaissance of nuclear power stimulates its global expansion, the international community needs to reconsider how to prevent disaster while increasing our reliance and stimulating the expansion to the far corners of the globe. In part, that can be accomplished through technological means or through other mechanisms that contain proliferation while permitting growth and stability.

Proliferation is a global concern. The IAEA somehow magically stands before us in this challenging era: no other international organization is held in such high regard, and assuring its continued viability is critical for future peace. Expanding its missions can provide greater assurance of peace and security in the future, provided those roles are considered carefully and implemented under arrangements that promote success.

Money will always be at the core of what the Agency can or should do in the future. While today the Agency relies almost exclusively on assessed contributions from national treasuries and from extrabudgetary contributions, most of which come from those same treasuries. Diversifying the financing arrangements can provide for growth, dropping the grueling debates on how growth could be financed to how the finance already attained can be best directed to secure sustainable economic development and international security.

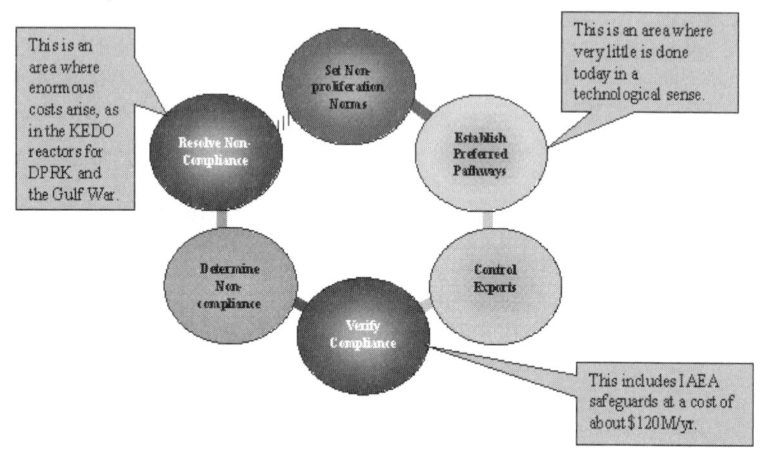

While IAEA safeguards are a critical part of this enterprise, it is, in fact, one with a rather small price tag. The other areas are in similar need, and the amounts needed may be substantially greater than what the IAEA could gainfully commit to enhanced verification.

ENDNOTES - CHAPTER 11

1. For the year ending December 31, 2005, the total amount expended on Nuclear Verification was $121,094,383.00, which includes disbursements and unliquidated obligations. For 2005, the assessed contributions for the IAEA totaled $316,473,124. GC(50)/8, The Agency's Accounts for 2005, p.54, p.112.

2. See Article XIV of the IAEA Statute.

3. The Geneva Group comprises the States that pay the bulk of the IAEA regular budget.

4. See GC(50)/8, The Agency's Accounts for 2005, p.112.

5. *Ibid.*, p. 113.

6. See INFCIRC/370, "Rules Regarding Voluntary Contributions to the Agency."

7. GOV/2005/86, IAEA Special Fund—Nobel Peace Prize for 2005, para. 6.

8. Available from *www.ocrwm.doe.gov/about/budget/index.shtml.*

9. Information from IAEA Power Reactor Information System, *http://www.iaea.org/programmes/a2/*

10. Energy Information Administration input.

11. On September 16, 1959, the World Bank made a loan equivalent to $40 million for the construction of a 150MWe [megawatt electric] GE BWR [boiling water reactor] at a site on the Garigliano River in Italy (Loan 0235). This was Italy's first nuclear power plant, and the Bank's loan financed almost two-thirds of the cost of construction. The plant began operation in 1964. In August 1978, it was shut down due to damage to one of the two secondary steam generators. In March 1982, the Italian Electricity Generating Board declared the plant to be out of service. Available from *web.worldbank.org/WBSITE/EXTERNAL/EXTABOUTUS/EXTARCHIVES/0,,contentMDK:20125474~pagePK:36726~piPK.*

12. Available from *psdblog.worldbank.org/psdblog/2006/04/go_nuclear_for_.html.*

ABOUT THE CONTRIBUTORS

THOMAS B. COCHRAN is Director of the Nuclear Program and holds the Wade Greene Chair for Nuclear Policy at the Natural Resources Defense Council (NRDC). He initiated NRDC's Nuclear Weapons Databook Project, and also a series of joint nuclear weapons verification projects with the Soviet Academy of Sciences. These include the Nuclear Test Ban Verification Project, which demonstrated the feasibility of utilizing seismic monitoring to verify a low-threshold test ban, and the Black Sea Experiment, which examined the utility of passive radiation detectors for verifying limits on sea-launched cruise missiles. He has served as a consultant to numerous government and nongovernment agencies on energy, nuclear nonproliferation and nuclear reactor matters. Dr. Cochran is a member of the Department of Energy's (DOE) Nuclear Energy Research Advisory Committee. Previously he served as a member of DOE's Environment Management Advisory Board; Fusion Energy Sciences Advisory Board and Energy Research Advisory Board; the Nuclear Regulatory Commission's Advisory Committee on the Clean Up of Three Mile Island; and the TMI Public Health Advisory Board. Dr. Cochran is the author of *The Liquid Metal Fast Breeder Reactor: An Environmental and Economic Critique* (1974); and coeditor/author of the *Nuclear Weapons Databook, Volume I: U.S. Nuclear Forces and Capabilities* (1984); *Volume II: U.S. Nuclear Warhead Production* (1987); *Volume III: U.S. Nuclear Warhead Facility Profiles* (1987); *Volume IV: Soviet Nuclear Weapons* (1989); and *Making the Russian Bomb: From Stalin to Yeltsin* (1995). In addition, he has published numerous articles and working papers, including those in *SIPRI Yearbook* chapters, *Arms Control Today*, and the *Bulletin of the Atomic Scientists*. He co-authored, with Dr. Robert S. Norris, the article on "Nuclear Weapons" in the 1990 printing of *The New Encyclopedia Britannica* (15th edition). Dr. Cochran received his Ph.D. in Physics from Vanderbilt University in 1967. He was Assistant Professor of Physics at the Naval Postgraduate School, Monterey, California, from 1967 to 1969; Modeling and Simulation Group Supervisor of the Litton Mellonics Division, Scientific Support Laboratory, Fort Ord, California, from 1969 to 1971; and from 1971 to 1973, he was a Senior Research Associate at Resources for the Future. He is the recipient of the American Physical Society's Szilard Award and the Federation of American Scientists' Public Service Award, both in 1987. As a consequence of his work, NRDC received the 1989 Scientific Freedom and Responsibility Award by the American Association for the Advancement of Science (AAAS). Dr. Cochran is a Fellow of the American Physical Society and the AAAS.

GARRY DILLON graduated in Applied Physics from the now University of Salford. After a initial career in radioisotope applications, he joined the UK Central Electricity Generating Board (CEGB) in 1963 and became the Health Physicist at Trawsfynydd Nuclear Power Station. In 1970, he transferred to CEGB London HQ, Nuclear Health and Safety Department, as nuclear safety compliance inspector for Bradwell and Sizewell Nuclear Power Stations. He joined the IAEA, Radiological Safety Section, in 1976 and was appointed Radiation Health and Safety Officer. He transferred to the Department of Safeguards (DSG) in 1978. In 1980, he established the IAEA's first fully functional Safeguards Field Office, located in Toronto, Canada, and served as head of that office until returning to

Vienna HQ in 1985. He served in a number of line-management positions within DSG, the last of which was as section head responsible for, inter alia, the verification of South Africa's voluntary abandonment of its nuclear weapons program. Mr. Dillon joined the Iraq Action Team in November 1993 as Deputy Leader (Operations) and was Action Team Leader from June 1997 until his retirement in October 1999. In mid-1994, he established the Nuclear Monitoring Group (NMG) whose principal function was the in-field implementation of the IAEA Ongoing Monitoring and Verification (OMV) plan—a critical component of which was a progressively introduced Wide Area Monitoring program. The NMG was continuously present in Iraq until its departure in December 1998. Throughout this time with the Iraq Action Team, he spent more than 350 days in-country and was actively involved in the carrying out of inspections, as well as interviewing Iraqi technologists and political officials.

JACK EDLOW graduated from George Washington University with a BA in Business Administration, then joined his father at Edlow International in 1969. He became Vice President of the company in 1970, then President in 1978. As President of Edlow International Company for over 20 years, Mr. Edlow has been responsible for all phases of the company's activities including transportation, warehousing, and logistic support services for nuclear suppliers and users worldwide. He has acted as sales representative for fuel cycle activities of leading international organizations involving supply of concentrates and conversion services and management of toll enrichment contracts. He has been active in consulting activities covering all aspects of the nuclear fuel cycle. Over a period of 3 years,

Mr. Edlow managed a special division of the Company established to provide special high security transport services to ERDA/DOE for sensitive materials. This system used a fleet of dedicated armored vehicles to move the materials across a nationwide network between various government and contractor facilities. Prior to the establishment of DOE's Office of Civilian Waste Management (OCRWM) program, Edlow International and the firm, Ridihalgh Eggers Assoc., teamed up to design a new generation of truck and rail transport casks. It has become the benchmark for many of the OCRWM financed and supported designs that have been subsequently developed. Mr. Edlow spent 2 years of his and the Company's effort and resources on convincing the Executive and Legislative Branches of the U.S. Government to support the Reduced Enrichment Research & Test Reactor (RERTR) program and to take back the high-enriched spent fuel discharged from such reactors. He has overseen the successful shipment of U.S. origin spent fuel from many research reactors in Europe and South America. Completed shipments have required the complex integration on all modes of transportation while supporting the DOE's 10-year program to complete shipment of material from four countries. Mr. Edlow is also Managing Director of Edlow International Australia Pty. Limited, Edlow International Company's Australian subsidiary in Melbourne. As President of Edlow East-West, Inc., Mr. Edlow has traveled to many Russian nuclear facilities and has acquired extensive personal knowledge.

PIERRE GOLDSCHMIDT is a visiting scholar at the Carnegie Endowment for International Peace and also a member of the Board of Directors for the Association Vinçotte Nuclear (AVN). AVN is a non-profit, authorized inspection organization charged with verifying compliance of nuclear power plants with Belgian safety regulations. Dr. Goldschmidt was the Deputy Director General, Head of the Department of Safeguards, at the International Atomic Energy Agency from 1999 to June 2005. The Department of Safeguards is responsible for verifying that nuclear material placed under safeguards is not diverted to nuclear weapons or other nuclear explosive devices and that there are no undeclared nuclear material or activities in non-nuclear weapons states party to the NPT. Before assuming this position, Dr. Goldschmidt was, for 12 years, General Manager of SYNATOM, the company responsible for the fuel supply and spent fuel management of seven Belgian nuclear plants that provide about 60 percent of the country's electricity. For 6 years, Dr. Goldschmidt was a member of the Directoire of EURODIF, the large French uranium enrichment company. He has headed numerous European and international committees, including as Chairman of the Uranium Institute in London and Chairman of the Advisory Committee of the EURATOM Supply Agency. Dr. Goldschmidt studied Electro-Mechanical Engineering and holds a Ph.D. in Applied Science from the University of Brussels; a Masters Degree in Nuclear Engineering from the University of California, Berkeley; and a B.A. in Electro-mechanical Engineering from the University of Brussels .

NIKOLAI NIKOLAEVICH KHLEBNIKOV is a national of the Russian Federation. He holds a Ph.D. in Chemical Technology from the State Research Institute of Rare Metals in Moscow. Dr. Khlebnikov started his career in 1970 as a researcher at the State Research Institute of Rare Metals where he worked for 8 years, followed by 8 years as a Section and

Laboratory Head in the Central Research Institute of Atomic Information in Moscow. He then worked for 8 years in the International Atomic Energy Agency (IAEA) in Vienna, first as a Section Head for System Studies in the Division of Concepts and Planning and then as a Section Head in the Division of Operations C responsible for safeguards implementation in European countries. He returned to the Ministry of the Russian Federation on Atomic Energy, Department for International Relations in Moscow where he worked as a Division Head and was responsible for nonproliferation issues and international organizations. He was a member of the Standing Advisory Group on Safeguards Implementation from 1994 until 1998. He joined the IAEA again in 1998, when he was appointed as Director of the Division of Technical Support with overall responsibility for the development and maintenance of equipment for verification of nuclear materials and training, a position he still holds. Dr. Khlebnikov has written about 30 publications in the area of chemical technology and about 25 publications in the area of safeguards.

EDWIN LYMAN is a Senior Staff Scientist in the Global Security Program at the Union of Concerned Scientists (UCS) in Washington, DC, a position he has held since May 2003. Before going to UCS, he worked at the Nuclear Control Institute for nearly 8 years, first as scientific director and then as president. He earned a doctorate in physics from Cornell University in 1992. From 1992 to 1995, he was a postdoctoral research associate at Princeton University's Center for Energy and Environmental Studies. Dr. Lyman's research focuses on security and environmental issues associated with the management of nuclear materials and the operation of nuclear power plants. He has published articles and letters in journals and magazines including *Science*, the *Bulletin of the Atomic Scientists*, and *Science and Global Security*. He is an active member of the Institute of Nuclear Materials Management. In the spring of 2001, he served on a Nuclear Regulatory Commission expert panel on the role and direction of the NRC Office of Nuclear Regulatory Research and briefed the Commission on his findings. In July 2001, he was again invited to a Commission meeting to discuss the licensing of new nuclear reactors in the United States.

QUENTIN MICHEL is Lecturer in Non-Proliferation and Sustainable Development at the Faculty of Law of Liège University (Belgium). He teaches also at the International School of Nuclear Law, OECD Nuclear Energy Agency and University of Montpellier 1, France. Dr. Michel is also an expert for the European Commission, the Belgium Federal Agency for Nuclear Control, and for the Belgian Government on weapons nonproliferation issues.

DAVIDE PARISE received his Ph.D. in Energy Management on Safeguards at Universita' "La Sapienza" (Roma) before joining the International Atomic Energy Agency (IAEA) in 2005 to develop software for the characterization of nuclear materials with gamma/X spectroscopy for safeguard use. In 2006 he joined the Novel Technologies Project as a safeguards system analyst to assist with the examination of candidate novel technologies that could be used by IAEA inspectors, mostly focusing on the use of laser technologies and on the detection of nuclear activities and facilities from a distance. In 2007 he joined the newly established Remote Monitoring Unit to design the remote monitoring infrastucture and to support IAEA inspectors in the field deployment of remote monitored surveillance and monitoring systems.

THOMAS E. SHEA was named Director for Defense Nuclear Nonproliferation Programs in January 2004 at the Pacific Northwest National Laboratory (PNNL) operated by Battelle Memorial Institute for the U.S. Department of Energy. PNNL's Defense Nuclear Nonproliferation Programs assist the National Nuclear Security Administration's Office of Defense Nuclear Nonproliferation in policy and technical activities aimed at preventing proliferation and nuclear terrorism, in nuclear safety, and in weapon-origin fissile material disposition. Prior to joining PNNL, he served for 24 years at the International Atomic Energy Agency (IAEA). At the IAEA, Dr. Shea helped to establish the basic IAEA safeguards implementation parameters and defined safeguards approaches for many complex nuclear facilities. He headed a section of inspectors for 11 years, responsible for safeguards implementation in Japan, India, Taiwan, Australia, and Indonesia. He established the Project Office for the JNFL Rokkasho Reprocessing Facility, and successfully headed a Tripartite Project with the Russian Federation and the People's Republic of China regarding safeguards at centrifuge enrichment plants equipped with Russian centrifuges. During the period from 1996 through 2003, Dr. Shea was Head of the IAEA Trilateral Initiative Office in the Department of Safeguards, responsible for program development and implementation activities associated with a possible new verification role for the IAEA: weapon-origin and other fissile material released from military applications. He also headed

IAEA activities related to a fissile material cutoff treaty, publishing a number of articles and briefing delegates to the UN Conference on Disarmament on six occasions. Dr. Shea was named to a UN Security Council Panel on disarmament in Iraq in 1999 and carried out an IAEA investigation of the technical requirements for the verification of the Comprehensive Nuclear Test Ban Treaty. He wrote the proliferation-resistance and physical protection parts of the U.S. Generation IV Roadmap and led the IAEA Safeguards departmental activities related to proliferation resistance. Dr. Shea was awarded a Special Fellowship from the U.S. Atomic Energy Commission. He received his M.S. in Nuclear Engineering and his Ph.D. in Nuclear Science from Rensselaer Polytechnic Institute. He is a Fellow of the Institute of Nuclear Materials Management.

HENRY D. SOKOLSKI is the Executive Director of the Nonproliferation Policy Education Center, a Washington-based nonprofit organization founded in 1994 to promote a better understanding of strategic weapons proliferation issues for academics, policy makers, and the media. He served from 1989 to 1993 as Deputy for Nonproliferation Policy in the Office of the Secretary of Defense under Paul Wolfowitz and received the Secretary of Defense's Medal for Outstanding Public Service. Prior to his appointment to this post, Mr. Sokolski worked in the Secretary's Office of Net Assessment on proliferation issues. In addition to his Executive Branch service, Mr. Sokolski served from 1984 through 1988 as Senior Military Legislative Aide to Senator Dan Quayle and as Special Assistant on Nuclear Energy Matters to Senator Gordon Humphrey from 1982 through 1983. He also served as a consultant on proliferation issues to the intelligence community's National Intelligence Council. After his work in the Pentagon, Mr. Sokolski received a congressional appointment to the Deutch Proliferation Commission, which completed its report in July 1999. He also served as a member of The Central Intelligence Agency's Senior Advisory Panel from 1995 to 1996. Mr. Sokolski has authored and edited a number of works on proliferation related issues including, *Best of Intentions: America's Campaign Against Strategic Weapons Proliferation* (2001), *Getting Ready for a Nuclear-ready Iran* (2005); *Checking Iran's Nuclear Ambitions* (2004); *Getting MAD: Nuclear Mutual Assured Destruction Its Origins and Practice* (2004); *Beyond Nunn-Lugar: Curbing the Next Wave of Weapons Proliferation Threats from Russia* (2002); *21st Century Weapons Proliferation: Are We Ready?* (2001); *Planning for a Peaceful Korea* (2001); *Prevailing in A Well Armed World* (2000), and *Fighting Proliferation* (1996). Mr. Sokolski has been a resident fellow at the National Institute for Public Policy, the Heritage Foundation, and the Hoover Institution. He currently serves as an adjunct professor at the Institute of World Politics in Washington and has taught courses at the University of Chicago, Rosary College, and Loyola University. Mr. Sokolski attended the University of Southern California and Pomona College and received his graduate education at the University of Chicago.

FRANK VON HIPPEL is Professor of Public and International Affairs, Woodrow Wilson School, and and co-chair of the International Panel on Fissile Materials. From September 1993 through 1994, he was on leave from Princeton as Assistant Director for National Security in the White House Office of Science and Technology Policy, and played a major role in developing U.S.-Russian cooperative programs to increase the security of Russian nuclear-weapon materials. In 2005 he chaired the American Physical Society's Panel on Physics and Public Affairs. He also chairs the editorial board of *Science & Global Security* and is a member of the editorial board of the *Bulletin of the Atomic Scientists.* Dr. von Hippel received his B.S. degree in physics from MIT in 1959 and D.Phil. in theoretical physics in 1962 from Oxford, where he was a Rhodes Scholar. During the following 10 years, while his research focus was in theoretical elementary-particle physics, he held research positions at the University of Chicago, Cornell University, and Argonne National Laboratory and served on the physics faculty of Stanford University. In 1974, his interests shifted to "public-policy physics." After spending a year as a Resident Fellow at the National Academy of Science, during which time he organized the American Physical Society's Study on Light-Water Reactor Safety, he was invited to join the research and in 1984 the teaching faculty of Princeton University. Dr. von Hippel has served on advisory panels to the Congressional Office of Technology Assessment, U.S. Department of Energy, National Science Foundation, and U.S. Nuclear Regulatory Commission, and on the boards of directors of the American Association for the Advancement of Science and the *Bulletin of the Atomic Scientists.* For many years he was the elected chairman of the Federation of American Scientists. Dr. von Hippel shared with Joel Primack the American Physical Society's 1977 Forum Award for Promoting the Understanding of the Relationship of Physics and Society for their book, *Advice and Dissent: Scientists in the Political Arena.* In 1989, he was awarded the Federation of American Scientists' Public Service

Award for serving as a "role model for the public interest scientist." In 1991, the American Institute of Physics published a volume of his selected works under the title *Citizen Scientist*, as one of the first three books in its "Masters of Physics" series. In 1993 he was awarded a 5-year MacArthur Prize fellowship. In 1994, he received the American Association for the Advancement of Sciences' Hilliard Roderick Prize for Excellence in Science, Arms Control, and International Security.

JULIAN WHICHELLO joined the Australian Nuclear Science and Technology Organisation (formerly the Australian Atomic Energy Commission) in 1973. From 1973 to 1987, he developed high-speed drive systems and electronic instrumentation for the Australian Gas Centrifuge Uranium Enrichment Project. As the Head of the Instrumentation Unit from 1983 to 1987, he collaborated in the development of a nuclear safeguards remote monitoring system for small scale enrichment plants and conducted investigations into the establishment of a stable isotope enrichment facility based on a combination of separation technologies (electromagnetic, laser, and vacuum arc centrifuge). Mr. Whichello was appointed to the International Atomic Energy Agency (IAEA) Surveillance Unit in Vienna in 1987. As Head of the Surveillance Unit from 2000 to 2005, he oversaw the development and implementation of a wide range of safeguards equipment and systems, including secure digital image surveillance and remote monitoring. Mr. Whichello is currently the Manager of the IAEA's Department of Safeguards Novel Technologies Project.

ROBERT ZARATE is a research fellow at the Nonproliferation Policy Education Center, a Washington, DC-based nonprofit organization founded in 1994 to promote a better understanding of strategic weapons proliferation issues among policymakers, scholars, and the media. He is concurrently researching and writing a book on the late American strategists, Albert and Roberta Wohlstetter. After graduating from the University of Chicago in 1999, Mr. Zarate worked from 2000 to late 2001 as a policy analyst at Steptoe and Johnson LLP in Washington, DC, focusing on international controls related to the import, export, and use of encryption and other dual-use items. In early 2002, he wrote for *Wired News*, covering the intersections of national security, technology, politics, law, and business. In late 2002, he returned to the University of Chicago to begin graduate studies. Mr. Zarate has published essays and articles in *The Weekly Standard, National Review Online, Wired News, E-Commerce Law Week*, and other periodicals.